Amor Mundi and Overcoming Modern World Alienation

Religion and Science as a Critical Discourse

Series Editors: Lisa Stenmark, San Jose State University and Whitney Bauman, Florida International University

Understanding religion and science as a critical discourse means building on theoretical issues and concerns to address social transformation, and issues of justice and global concerns. Contributions to this series will employ multiple perspectives upon the process of doing and thinking "science and religion" together, but ultimately see the relationship of religion and science as creating space for a kind of critical discourse. This might mean: exploring disagreements between two authoritative disciplines that challenge one another; incorporating critical theories and discourses (understood narrowly as the Frankfort School, and, more broadly, as critical race theory and feminist, postcolonial, and queer approaches within the social sciences, natural sciences, or humanities); or a focus on voices from outside the dominant discourse, which in the case of this series means people from outside of the western academy. In each case, the goal is to shake up assumptions, challenge givens, and open up space for new questions and new perspectives so we can think about pressing problems in a more productive and inclusive way.

Recent Titles in Series

Amor Mundi *and Overcoming Modern World Alienation*, by Justin Pack

Navigating Post-Truth and Alternative Facts: Religion and Science as Political Theology, edited by Jennifer Baldwin

Unsettling Science and Religion: Contributions and Questions from Queer Studies, edited by Lisa Stenmark and Whitney Bauman

Amor Mundi and Overcoming Modern World Alienation

Justin Pack

LEXINGTON BOOKS
Lanham • Boulder • New York • London

Published by Lexington Books
An imprint of The Rowman & Littlefield Publishing Group, Inc.
4501 Forbes Boulevard, Suite 200, Lanham, Maryland 20706
www.rowman.com

6 Tinworth Street, London SE11 5AL, United Kingdom

Chapter No. 1, 2, 3, 4, 6. The Human Condition by Hannah Arendt, nonfiction source. Arendt, Hannah. The Human Condition. (Chicago: University of Chicago Press, 1998).

British Library Cataloguing in Publication Information Available

Library of Congress Cataloging-in-Publication Data Available

ISBN: 978-1-4985-9134-8 (cloth)
ISBN: 978-1-4985-9135-5 (electronic)

Contents

Series Preface

It has been fifty years since Ian Barbour wrote *Issues in Science and Religion* (HarperCollins 1971), a publication that marks the beginning of what has become the broad interdisciplinary field of science and religion. But despite interest from a wide range of disciplines and a concern for a wide range of issues, the field continues to be marked by a concern for theoretical issues, particularly those that focus on the truth claims of religion and science. One result is that the discourse(s) of religion and science often lacks what one editor (Lisa) calls a "prophetic voice" and another (Whitney) a planetary perspective. This is a matter of some concern because there is a pressing worldwide need to address serious issues that exist at the intersection of western sciences, technology and religion, and there continue to be (serious) concerns about the status of scientific claims, and the totalizing tendency of scientific claims over and against religions and other knowledge systems. There remain ongoing concerns because the discourses involved in the western academic science and religion discourse (SRD) have been historically complicit in creating many of the problems that we need to address.

Religion and science needs critical engagement, yet it is largely lacking largely because an emphasis on truth claims seeks to downplay differences, which is the entry point for critical engagement. This discourse lacks a (self) critical perspective, and this series attempts to address it through a somewhat fuzzy use of the idea of critical discourses. Critical discourses and theories can, of course, refer to Critical Theory in the narrow sense (The Frankfort School—Max Horkheimer, Theodor Adorno, Jürgen Habermas—and their intellectual descendants), but it can also be used in a broader sense, to refer to feminist, critical race theory, postcolonial, queer approaches within social sciences, and also legal theory and literary approaches. All of these discourses help to challenge dominant understandings of the world in order that multiple perspectives and experiences might be heard. Religion and science as a critical discourse can also mean avoiding a doctrines and discoveries approach—which emphasizes the truth claims of religion and science—and exploring areas of disagreement as a way of opening up the science and religion discourses to more voices, and more perspectives. By critical discourse, we mean all of these, and more, because none of these approaches is sufficient, but all of them are crucial for thinking about the planetary community and our moral and ethical responsibilities to human and earth others.

Not every criticism is critical, of course, and what all of these approaches have in common is that they engage in criticism with a purpose: because theory must challenge *and* transform human structures—not just for criticisms sake, but to create a world more attune to human—and planetary—flourishing. These approaches are suspicious of absolutes whether scientific or religious or otherwise. They understand all knowledge as historical and political—as shaped by human culture and human interests—and they are critical in the sense that they confront social, historical, and ideological structures within culture and society in the interest of justice and flourishing, but also as a way to open up space for sound judgments. This is criticism aimed at judgment in the sense that political philosopher Hannah Arendt understood it: the ability to think what we can do. Thus, a critical discourse involves a many-sided engagement with the human and natural worlds, including multiple social and cultural locations (most particularly non-Western perspectives). It is interdisciplinary (humanities, social sciences, natural sciences), and it is connected to action, not merely deciding what is the case, but also how to act in response. A critical discourse does not merely point out the problem—or complain about it—it suggests solutions and represents a call to action. Ultimately, this is all meant to promote an approach to the intersection of religion and science that is not a discipline in the conventional sense of the word, but an approach oriented towards social transformation, justice and global issues.

This third book in the series approaches the conversation from the point of view of philosophy, as the author argues that as a result of Modernity and a modern scientific worldview, we have forgotten the pre-modern experience, and coexisting alternatives experiences outside of Modernity, of "love of the world." The result of this loss is a diminished understanding of the concept of love as well as the idea of the world. This loss is felt spiritually, philosophically and scientifically, and has led to existential alienation and has helped to create contemporary environmental crises, such as climate change. As a contribution to developing just such an understanding, he draws upon two disparate understandings of the concept "love of the world" —Hannah Arendt's concept of *amor mundi as* love of a shared public world, and Vine Deloria and Daniel Wildcat in terms of loving the more than human world. In a sense, Justin Pack is arguing that we need to replace the cold "Anthropocene" with a (re)new(ed) understanding of "Amor Mundi" that integrates philosophy, along with the sciences and the humanities in general under the common, earthly politics of acting in ways that once again show love for the world. His contribution helps us work our way towards this renewed sense of loving the world. In addition, he broadens the science and religion discourse beyond its over-emphasis on western traditions, drawing on indigenous perspectives and, finally, contributing to a consideration of the "political" impact of our discourse through inquiring what love of the natural world hast to do with our public life and vice a versa.

Introduction

This book concerns love of the world. Both "love" and "world" deserve careful attention separately, but especially together as the phenomenon of love of the world—*amor mundi*. In this book, I will be examining two forms of *amor mundi* in premodern Western thought and a contemporary non-Western form of *amor mundi* and claiming that despite love of the world being a fundamental concern in many pre- and nonmodern cultures and traditions, today neither love, nor world, nor love of the world (least of all) play a prominent role in contemporary Western thinking. The idea that love is fundamental to knowledge and that we should love the world is largely forgotten and misunderstood now. The modern scientific worldview has so radically changed what the world is understood to be and how we relate to it that it is difficult to comprehend premodern ideals of loving the world.

It is not hard to see this modern world alienation all around us. For example, I do an exercise with my undergraduate students sometimes in which I state a word and ask them to shout out the first thing that comes to mind. When I do this with the word "universe" their answers are telling. They respond with words like: "cold," "huge," "empty," "expanse," "physical," and so on. Rarely, a student will shout out "home!" The contrast here is remarkable—in fact, quite often they stress that the universe is NOT our home, that we live on the "third rock from the sun," that life is an "accident" and that there is no meaning in the universe. When I point out how the language they are invoking is, in many ways, inhuman and contrast it with the idea of home, they have various reactions: some students are struck by this contrast and feel sad, some dismiss the need for home as religious and antiquated, some take a sort of stoic pride in the "truth" of a cold universe. But when I ask them to put aside facts and think about what it would be like to (mistakenly or not) live in a cosmos that is a home for humans, they can see how losing the cosmos for a universe could be felt as a profound loss.

If do this same exercise with the word "love" the responses tend to be words like "sex," "passion," "emotion," "relationship," "connection," "family," and "friends." Sometimes students answer "God" and occasionally hobbies or other things they like that bring them meaning. I don't think I have ever had a student respond with "world."

Love in contemporary consumerist society tends to be sexualized, romanticized, and individualized. Love is primarily understood as something that occurs between romantic and sexual partners. The history of love can trace the sexualization of love to the rise of advertising in the early twentieth century.[1] Sex sells, and we are inundated with eroticized imagery that plays on romantic and sexual desires, promising fulfillment if we shop smartly. We see this now in our everyday language: "making love" refers to sex, not building relationships with others.

Historical and cultural analysis of love shows that this is a rather narrow and superficial understanding of love. There is good reason to lament how this sexualized love can obscure and warp our understanding of other kinds of love. Learning itself was treated as an activity of love and desire by Plato, and the relation of teachers and students can be an exciting and charged loving relation,[2] but when I ask students what is the relationship of love to education, they find this question extremely awkward at first. They connect love and sex so strongly that they immediately deny love (sex) has anything to do with education. When they get over this initial reaction, they recognize that love sometimes plays a role in education—many have had teachers or courses that they loved and these experiences have often had a profound effect on them. But they also tend to think that education is about being objective and searching for facts. Love is too emotional—perhaps it is good to get students excited, but in serious, scientific study, love is to be avoided.

The kind of love I discuss in this book is radically different from the atomistic, sexualized love of modern consumerism. The term *amor mundi*, love of the world, comes from Hannah Arendt. I will go into more detail on her specific definition, but I want to expand her phrase to appreciate the ambiguity of the "world" part of *amor mundi* and recognize that there are different kinds of love of the world. For Arendt, she means by "world" both the physical objects (both mundane and artistic/architectural) and the social norms, narratives, and meanings that create a human "world" that most humans are born into. But there are other important understandings of what *amor mundi* is and are less anthropocentric. In the Native American thought of Vine Deloria and Daniel Wildcat, there is also a remarkable sense of *amor mundi*, but "world" is taken to be all the plants, animals, soils, rocks, rivers, and landscapes that are related and interact in a particular place. And there is another form I will examine: according to Rémi Braque, Plato's later philosophy encourages an *amor mundi* that extended to the cosmos—a kind of cosmic love.

What a rich experience we have forgotten! In the three traditions, I will examine some kind of love of the world tied humans to the cosmos (Plato), the Earth/Nature/place (Deloria and Wildcat), and/or a city/culture (Arendt), but even a cursory look at anthropology reveals myriad complex relationships with the world, many of which could be construed as kind of love of the world. Obviously, what exactly this looked like varies radically in different traditions, but my focus on love of the cosmos (Plato), love of a political/cultural world (Arendt) and love of the Earth/Nature/place (Deloria and Wildcat) serves a couple of purposes: (1) to show how different forms of *amor mundi* can be while still tying together people to some kind of greater whole (world); (2) to give a historical account of the decline of *amor mundi* with the rise of science and modernity in the West; (3) to give an indication of what is lost in this shift (cosmos, connect with place and nature, connection with community and culture); (4) to introduce the environmental, ethical, and social implications of the loss of *amor mundi*, including modern alienation; and (5) to encourage regaining a love of the world.

I treat this as not merely an academic task, but an existential one. By this I mean that thinking through *amor mundi* is not a matter of observing and recording interesting traditions, but of opening ourselves up to ways of relating to the world around us. We need this connection to the world. As we will see, Arendt claims that love of a political/cultural world is vital to democracy and public happiness. Seeking only private happiness will never leave us fulfilled—we need public happiness. Additionally, Deloria and Wildcat show us that love of the Earth and Nature is vital to the Earth, the environment, and to our relationship with them. In light of the deepening environmental crisis in which we are seeing massive species die off, global warming, the pollution of the ocean, and the ongoing destruction of forests, it is hard to understate how alienated we have become from the Earth and how much the Earth itself needs us to reverse our course. *Amor mundi* is something we humans need and something the Earth needs.

IS PHILOSOPHY UP FOR SUCH A TASK?

Philosophy is the only academic discipline that includes the word "love" in its own name. Most other disciplines study, philosophy supposedly loves. With that said, most people do not think of philosophers as lovers—neither in love, nor infatuated, nor full of love, nor loving. Perhaps the poets, the artists or theatre folks, but philosophers tend to be like other academics: dry, rational, and often quietly productive.

Every student who has taken an existentialism course knows philosophy doesn't have to be this way. Thinkers like Kierkegaard and Nietzsche excoriated the petty, calculative thinking of their times. Nietzsche called for a joyous science to replace state-serving menial positivism:

> All great problems demand *great love*, and of that only strong, round, secure spirits who have a firm grip on themselves are capable. It makes the most telling difference whether a thinker has a personal relationship to his problems and finds in them his destiny, his distress and his greatest happiness, or an "impersonal" one, meaning that he can do no better than to touch them and grasp them with the antennae of cold, curious thought.[3]

My hope is that this book will introduce *amor mundi* as a living, existential task. But it is an existential task that connects the individual up with others and with the world. Existentialism was criticized by thinkers like Heidegger and Arendt for being too individualistic, too focused on individual meaning and individual responsibility. Additionally, existentialism has a tendency to be world weary, despairing, and pessimistic. *Amor mundi*, as a task, connects the individual up to others and the world. It retains a focus on responsibility, but sees meaning as something that is out in the world, not created by each of us in our own heads.[4]

Hans Jonas, a student of Heidegger, calls for an "existential interpretation of biological facts"—by which he means the expansion of existential concerns for meaning and responsibility to the entire living realm.[5] This means both recognizing that nonhuman plants and animals have concerns about their existence but also that human existence is tied to these beings and hence existential concern must go beyond the individual. *Amor mundi*, as we will see, seeks to do the work Jonas is calling for, but also points us even beyond living beings and connects us to the Earth and potentially the cosmos.

This book, then, is attempting to philosophically think what it would mean to reconnect with and love the world. Speaking in terms of plants, Robin Wall Kimmerer argues that

> the sanitized suburban life has succeeded in separating us from the plants that sustain us. Their roles are camouflaged under layers of marketing and technology. You can't hear the rustle of corn leaves in a box of Foot Loops. Most people have lost the ability to read the role of a medicine plant from the landscape and read instead the 'directions for use' on a temper-proof bottle of Echinacea. Who would recognize those purples blossoms in this disguise? We don't even know their names anymore. The average person knows less than a dozen plants, and this included such categories as "Christmas Trees." Losing their names is a step in losing respect. Knowing their names is the first step in regaining our connection.[6]

Referring to her Native American heritage she continues:

> Our ancient teachers tell us that the role of human beings is respect and stewardship. Our responsibility is to care for the plants and all the land in a way that honors life. We are taught that using a plant shows respect for its nature, and we use it in a way that allows it to continue bring its gifts. The role of our sacred sage is to make thoughts visible to the Creator. We can learn from the teacher and live in such a way that our thoughts of respect and gratitude are also made visible to the world.[7]

Kimmerer's concern about the deep disconnection between contemporary humanity and the world has been made over and over again both from within the Western tradition and in response to it from non-Western traditions—from the Romantics, from the Transcendentalists, from Continental philosophers, from environmentalists, from pop culture, from Native Americans, from Buddhism, and so on:

> This great lie we live [that there is a disconnected ego and that it can be self-sufficient] lulls us into a religious, ethical, and existential slumber; and we are now living out its very real consequences with our nearly global inability to realize our interconnectedness to the world and its creatures. This lack of interconnectedness manifests in our failure to design human habitats and food production systems in equitable and meaningful ways, to use land and other resources responsibly and justly, to build human communities that integrate with the planet's macroclimate and microclimates, and to live harmoniously with each other. Nietzsche and Nishitani challenge us not to just focus on the elements of the world themselves such as plants, animals, rocks, soils, and water as if they are somehow out there, but rather to assemble ourselves within the matrix of multitudinous relationships created by those plants, animals, rocks, soils and water. To become synergetic with the various forces of nature is to learn how to mimic emerging patterns as enhanced by the interactions of those elements found in nature. To accomplish all of this, the ego that underlies our sense of self must be abolished (Nietzsche) or broken through (Nishitani).[8]

These diverse and varied criticisms are clear that at the center of these problems is our disconnection and alienation from each other, from nature and from the Earth. In our consumeristic atomism we are destroying the very things that make our existence possible. We have lost *amor mundi* and we need to regain it.

Despite there being many different kinds of *amor mundi*, in this text I will examine three: love of the cosmos, love of the human world, and love of Earth/Nature/place. All three have been diminished or forgotten as a result of the rise of modern science and modernity in general. I will use the first, love of the cosmos, to tell the historical story of this shift in chapters 1 and 2. I do not claim that we can recover love of the cosmos as articulated by Plato, as it required a certain metaphysical understanding of the cosmos—one that

Copernicus undermined. Platonic cosmic love is important to understand, however, both because it shows the prominence of *amor mundi* in Western thought and because it is such a radical foil to our modern understanding. When we understand the metaphysical structure of the Timaean cosmos and how this may have shaped premodern Western epistemology, ethics, and human anthropology, we also can see more clearly the significance of the break with this that results from modern science and the how radically different the modern "universe" is from the premodern "kosmos."

The other two forms of *amor mundi*, love of the human world and love of the Earth/Nature/place are the topics of chapters 3 to 5. I use Arendt for the former and Deloria and Wildcat for the latter. While I do not think it possible to readopt Timaean metaphysics, I will suggest it is possible to recover *amor mundi* in both of the sense outlines by Arendt and these Native American thinkers. I claim we need both of these forms of love as they are pivotal to democracy, public happiness, the Earth, and the environment.

The topic of *amor mundi* stands at a critical nexus of issues and is related to a fascinating variety of topics. I have already claimed that *amor mundi* is oddly neglected in the philosophy of love and existentialism and that both would be much richer by recognizing the critical role it should be playing in both of these traditions. In addition to the philosophy of love and existentialism, *amor mundi* is of deep relevance to environmentalism. In fact, it should be clear that overcoming our modern world alienation and learning to love the world is so vitally important to recovering better relationships with the Earth and nature that I do not feel the need to present a systematic overview of environmental philosophy (as much as I like it) nor argue for the place of this topic in it. This book is a direct response to the current environmental crisis.

This book is also a response to what could be called naive Enlightenment optimism. This is the attitude that we see in thinkers like Steven Pinker who claim, against culture watchers and worriers, that modern science and modern reason have drastically improved life for humans to such a degree that we can assert life is better than ever.[9] In thinkers who feel this way, I find there is often a tendency to demonize the past and romanticize the present—finding comfort in measures like GDP and life expectancy while ignoring questions of meaning, purpose, exploitation, and environmental destruction. I have found that skepticism about modern progress is often dismissed as ignorance about how bad the past was—either in terms of violence, freedom, or quality of life. But I would claim the historical ignorance goes the other way and that native Enlightenment optimists don't understand how modern science has led to the death of the cosmos (what Nietzsche called the death of god), the subjectification of values (morality is no longer out in the world but in our heads), the scale and scope of environmental destruction, and the alienation of modern atomism. To worry about these shifts is not to want to abandon modern medicine and go back to the 1400s, but to recognize that the issue is far more complex than naive optimism makes it out to be. There are reasons premodern and nonmodern peoples often resisted and continue to resist modern colonialism: despite the many benefits of modern society they correctly understood and understand that there are critical flaws to modernity. By the end of this book, I will claim that modern atomism is crude, absurd, and destructive. It does not lead to a society filled with *amor mundi*, but a society of masturbational alienation.

As such, there is also a strong focus in places in this text on the philosophy of science. Unfortunately, criticism of modern science is often superficially dismissed as coming from a reactionary conservativism. Science is often erroneously treated as an objective procedure, a supposedly neutral tool that has no inherent morality. Twentieth-century philosophy of science has thoroughly debunked this claim and has shown that science cannot simply be divorced from the cultural context that it functions within, nor from the social and cultural history that has informed it, nor from ethical values. Science is not done by robots, but humans with values, purposes, and funding issues. Nonetheless, there is a common faith in science among Enlightenment optimists that too often fails to recognize the long and disturbing history of science and "scientific" intervention into society. At worst, this faith functions as a scientism (a related kind of faith can also manifest with regard to economics as economism) that claims scientific knowledge is superior to all other forms of knowledge. This attitude can lead to a dangerous dismissal of morality and other nonscientific questions of meaning and purpose.

This book should also be of interest to those who are interested in religion and spirituality, phenomenology, and Native American philosophy. Lastly and perhaps most narrowly, for Continental philosophers and scholars of Arendt, I am presenting her as a thinker of love and exploring her concept of world in great detail. While there is good work that has been done on each of these separately, there is surprisingly little written on her understanding of *amor mundi* itself.

Chapter 1 discusses the model of the cosmos and the kind of love it inspired that was introduced by Plato in the *Timaeus* and remain a popular metaphysical and ethical worldview the premodern Western philosophical understanding until the Copernican revolution. Following Rémi Brague, this chapter describes the premodern ontology of the cosmos, shows how morality was objectively in the cosmos itself and how the cosmos functioned as a meaningful home for humanity. In this context, to become a moral person entailed loving and imitating the cosmos. This chapter also uses Nussbaum to examine what Plato meant by love and how the trajectory in his writing moves from the search for scientific objectivity in his early writings to love as the key to knowledge in his later works. This chapter thus introduces love, world, and love of the world.

Chapter 2 examines how modern science undermined the Timaean cosmos, leaving us with a radically different understanding of the world around us: instead of the living cosmos, the mechanical universe. This, of course, comes with an epistemology that rejects love in favor of "objective" scientific methodology and leaves us in a very precarious situation with regard to ethics. I present three arguments about the implications of this shift: (1) Merchant's claim this leads to "the Death of Nature," (2) Nietzsche's claim this leads to the death of God, and (3) Hans Jonas's claim that modern thought occurs under the "ontological dominion of death." I finish with a discussion of Nietzsche's attempt to overcome this through *amor fati*.

Chapter 3 explains what Arendt means by "world." It traces how a concept that Arendt claims was so fundamental to the ancient Greeks became underappreciated and largely lost. Arendt never directly discusses love of the world, but *The Human Condition* points to it indirectly by tracing how it was lost.

Chapter 4 reconstructs what Arendt means by *amor mundi* by connecting it to her larger body of work, including her defense of Socratic (friendly) dialogue against Plato's (early) scientific quest for truth and her later discussion of revolutions as attempts to recover the kind of freedom and public happiness accomplished in the polis.

Chapter 5 turns to Native American philosophy to push back against Arendt and encourage a more expansive understanding of who is included in *amor mundi*. According to Deloria and Wildcat, Native American ontology requires love not be limited to humans or the human created world, but extended to all the peoples (plants, animals, rivers, etc.) of a particular place. From this perspective *amor mundi* refers to love of the Earth.

Chapter 6 ties these discussions to our contemporary social situation by discussing how *amor mundi* challenges both modern (sexualized, romanticized, and individualized) love and modern atomism. This task is more complicated that it looks because of how what Charles Taylor calls the "modern hydra" is deeply entrenched and tied to our contemporary understanding of the self, society, and freedom.

The conclusion argues that *amor mundi* is integral to overcoming our modern world alienation(s), creating healthy democratic practices, public happiness, and saving the Earth. I give some brief suggestions for how to foster *amor mundi* and discuss how these different forms of *amor mundi* help to support a politics of Nature that seeks to give Nature a voice in politics.

NOTES

1. Seidman, Steven. *Romantic Longings: Love in America, 1830–1980* (New York: Routledge, 1993); Ewen, Stewart. *Captains of Consciousness: Advertising and the Social Roots of the Consumer Culture* (New York: McGraw-Hill, 1976).

2. hooks, bel. *Teaching to Transgress: Education as the Practice of Freedom* (New York: Routledge, 1994).

3. Nietzsche, Friedrich. *The Gay Science* (New York, NY: Vintage, 1974), 345.

4. Julian Young claims that this was one of Heidegger's fundamental concerns. See Young, Julian. *The Death of God and the Meaning of Life* (New York: Routledge, 2014).

5. Jonas, Hans. *The Phenomenon of Life: Toward a Philosophical Biology* (Evanston, IL: Northwestern University Press, 2001), xxiii.

6. Wall Kimmerer, Robin. *Gathering Moss: A Natural and Cultural History of Mosses* (Corvallis, OR: Oregon State University Press, 2003), 101–102.

7. Wall Kimmerer, *Gathering Moss*, 110.

8. Jones, David. "Empty Soul, Empty World: Nietzsche and Nishitani," in Bret W. Davis, Brian Schroeer and Jason Wirth, eds. *Japanese and Continental Philosophy: Conversations with the Kyoto School* (Bloomington, IN: Indiana University Press, 2011), 105.

9. Pinker, Steven. *Enlightenment Now: The Case for Reason, Science, Humanism and Progress* (New York: Viking, 2018).

Chapter 1

Cosmic Love or Loving the Living Cosmos

What would it be like to love the world? Really *love* it. To look up at the stars and see moral perfection? To look out the window and wonder at the glory of this world? Pop culture tells us to love life: "you only live once" and "carpe diem!" Of course, this mostly means we should spend money on the weekends. Despite all the noise telling us to live life to the fullest, if we put this in terms of loving the world, there is surprisingly little philosophical analysis of what this means—especially considering that (cheap advertising soundbites aside) loving the world around us does seem to be something fundamentally and existentially important.

This appears even odder when we take into account that love of the world may have been a primary concern in premodern western philosophy. If this is indeed the case, then the absence of love of the world in contemporary thought and action should lead us to ask two questions: first, what is this love of the world that was apparently such a primary concern in western thought previous to modernity and, second, how have we forgotten it? This chapter will deal with the first question. The next chapter deals with the second.

What it means to love the world means something different depending on what is meant by "world." Over the course of this book, I will be examining three structurally different kinds of *amor mundi*, each with a different understanding of what is meant by "world." Briefly, the three are love of the cosmos, love of a political/cultural world (Arendt), and love of the Earth/Nature/place (Native American philosophy). While I will focus on the latter two forms of love of the world in chapters 3 to 5, my focus in this chapter is love of the cosmos. I go into greater detail with the latter two forms of love of the world because I will argue there are important and recoverable ethical, social, and existential implications that follow from them. Love of the cosmos also has radical implications, but the Timaean model of the cosmos that underlies premodern western *amor mundi* is not recoverable. In other words, while it is important to understand the historical significance of love of the cosmos in order to understand how modernity has changed us, I do not think we can recover it in a way that makes it a live option—at least in this Platonic form. *Amor mundi* in the other two senses I will examine, however, may be recoverable.

Thus, my focus on this chapter is in one sense historical. I want to remember the traditional importance of *amor mundi* and emphasize that this task was considered by some thinkers as one of the highest joys and primary concerns in premodern western experience. This is an act of *remembrance*, as modern science has undermined the Timaean cosmos and replaced it with a mechanical universe, and this shift has had such a profound effect on our ontology, epistemology, and ethics that most have forgotten *amor mundi*. As such it is not just a historical exercise, but opening up a series of questions and issues that are and should be in the forefront of human concerns. As I mentioned in the introduction, I take this to mean that *amor mundi* is an existential issue and that attempting to understand the history of *amor mundi* in Western thought can help us better understand our condition here and now, live more richly, more carefully, and more in touch with nature.

Why begin the approach to *amor mundi* from the Western tradition? First, because it is the Western tradition that has undermined *amor mundi*. To understand the modern decline of *amor mundi* means to tell a story that plays out in Western thought. This does not mean that there are non-Western societies that do not have some version of *amor mundi*. Indeed, there are many, and we will examine one such non-Western example later (Native American thought). In fact, to be clear, I think it is critically important that Western philosophy and Western thinkers in general be more attentive to alternative and/or nonmodern ontological, epistemological, and ethical traditions. By being more aware of the diversity of traditions thinkers raised in the Western tradition can become more aware of its limitations and potential flaws. Unfortunately overconfidence in Western methods and intellectual narrowness often leads to ignorance of alternatives and excessive faith in the status quo. If I am correct, then anthropology, history, ethnic studies, and feminism are disciplines and traditions that should be more important than they currently are in academia.

My purpose in this chapter, then, is to explain love of the cosmos as a pivotal task of premodern Western philosophy and attempt to show the questions this analysis opens for us about contemporary practices, ethics, and self-understanding. I will approach this in three parts. First, I will use Rémi Brague's *Wisdom of the World* to examine the ontological, anthropological, and ethical implications of Plato's Timaean cosmos—the model of the cosmos that Brague claims dominated Western premodernity until the Copernican revolution. Second, to clarify this connection between humans and cosmos, I use Charles Taylor to discuss how premodern humanity was a "porous" self as opposed to the "buffered" self of the modern individual. While there are many radical changes that accompany the shift away from the premodern cosmos to the modern universe, understanding this particular feature is important to understanding the intimate connection for some thinkers between human and world (cosmos) in premodern Western thought. Third, I want to finish the chapter by discussing how love was pivotal to knowing and relating to the cosmos despite a deep ambivalence in premodern Western thought about love. I will use Martha Nussbaum's discussion of how Plato moved away from a protoscientific approach in his early work to love in his later works to show some of the tensions involved.

THE TIMAEAN COSMOS: METAPHYSICS

There are a variety of texts that attempt to trace the shift I am tracking in this chapter, including Alexandre Koyre's classic *From the Closed World to the Infinite Universe* and, in a different way, Charles Taylor's wonderful *A Secular Age*. I will follow primarily Rémi Brague's *The Wisdom of the World* however, because it does an excellent job of connecting the metaphysical structure of the cosmos to the ethical and anthropological implications of such a cosmos. Understanding that premodern ethics in this model was intimately tied to premodern physics is necessary to understand why the modern divorce of these two has led to a troubling situation for modern ethics.

Brague argues the ancient world had multiple competing visions of the cosmos and recognizes that the "standard" model he reconstructs is not one that held sway over every individual from Plato to Copernicus. In fact, Brague presents four ancient models of the cosmos: the Timaean, Atomist, Scriptural, and Gnostic. In addition to these four competing models, there were idiosyncratic figures like Socrates who manifested a disregard for the cosmos that made him a true oddity in the ancient world (Brague uses the oddness of Socrates' disinterest in the cosmos as evidence of how important the cosmos was for human activity in the ancient world).[1] Brague's claim is that the model of the cosmos presented by Plato in the Timaeus eventually won out, was widely accepted and, with the prestige of having been authored by Plato, served as the "standard vision of the world" for most intellectuals until the Copernican revolution. This included Christians, Muslims, and Jews who combined the Timaean model with scripture.[2] This model had at times and in different cultures quite different variations, and yet Brague claims there is enough commonality to recognize a shared heritage. Note that Brague's claim is that this was the "standard vision of the world" for most intellectuals who tapped into the western philosophical tradition—commoners and folk traditions were often radically different.

Let me note quickly, that one does not have to accept the breadth of Brague's claim (that the Timaean cosmos was the "standard model" of the cosmos in the premodern West among intellectuals) to see the value of his analysis. In other words, even if he overreaches on his claims of the general acceptance of this model in the intellectual classes, he nonetheless points us to a very different understanding of the cosmos, ethics, and human nature that in all its potential variations is strange enough from the modern perspective that it merits careful elucidation. While clearly there was not one model that ruled all people in the West throughout premodernity (which Brague clearly admits), to use Weberian terms, Brague constructs a kind of ideal type, a model that helps us to understands the many variants, their similarities, and their differences. For my purposes here, we don't need to delve into all these variations, so the presentation I offer will likely feel more monolithic than the one Brague presents.

So what was the Timaean cosmos like? According to Brague, it was like a large onion composed of different layers, with the Earth in the center.[3] From the Earth at the center of the onion, the layers of the onion form spheres around the Earth. Instead of having elliptical orbits, the moon, the stars, the planets, and the sun are seen as being affixed to spheres that rotate around the earth. This may sound strange from a modern scientific perspective, but if you go outside and watch the stars at night, taking the time to observe how they are moving over a period of hours, it does look like they are moving in an arcing pattern across the sky—and you can see why it would make sense to posit that they are affixed to a sphere that is moving around the Earth, like being in a planetarium. Entire constellations move in gentle curves across the night sky. Of course, other objects, like the moon or the planets (Venus being the most visible), move in different, sometimes irregular, arcs. They must, therefore, be moving on different spheres. Thus the earth is nested in a series of spheres that are moving at different rates around it.

So, as different as it appears from the modern understanding of the universe, there is an intuitive sense to the idea that the Earth is at the center and that everything else is rotating around the Earth on spheres—after all, if the Earth were moving, wouldn't we all fly off? (Thus, the much later debate about the existence or nonexistence of an aether). As Arendt points out, premodern physics is based on everyday experience, while many of the discoveries of modern science are counterintuitive and often fly in the face of common sense (like a feather and a bowling ball falling at the same rate in a vacuum).[4]

According to the *Timaeaus*, there is one particular sphere that is very important, and this is the nearest sphere, the sphere of the moon. Everything below the sphere of the moon, including the Earth, is qualitatively different from everything above the sphere of the moon. Below the moon is the realm of Becoming—where everything is made of the elements and change is visible everywhere.[5] Above the moon is the realm of Being, where movement is constant, regular, predictable, orderly, and eternal.

Like the idea of spheres, the idea that the Earth and everything below the moon is the realm of Becoming and change makes a certain level of sense. We can look around at the world around us and see plants, animals and humans moving, eating, fighting, getting older, and changing. Even landscapes are slowly and inevitably changing. The stars, on the other hand, manifest seemingly eternal patterns and regularity. Their consistency allows sailors to orient themselves. Thus, when comparing the stars to the Earth, the Earth can appear quite chaotic.

On Brague's reading, Plato gives the chaos and constant change of this realm a negative evaluation in comparison with the stars: The realm of Becoming is one of imperfection, while the stars above us move in perfect, eternal, unchangeable patterns. And here the ethical and anthropological implications of this begin to come to the fore.

THE TIMAEAN COSMOS: ETHICAL AND ANTHROPOLOGICAL IMPLICATIONS

The realm of Becoming below the moon is not evil, but mixed good and evil.[6] This fits human experience which is full of great joys, small pleasures, great evils, and small pains. Human history is one of wars, killing, and madness but also love, friendship, and happiness. It is all here, whirling and chaotic, rarely predictable and never finished in this realm of change.

The realm of Being above the moon is, unlike our mixed state of becoming, perfect, good, and beautiful. The cosmos, then, is a realm of mixed good and evil below the moon, surrounded by a realm of perfection and beauty, like an egg yolk surrounded by the egg white. According to Brague, it follows from this structure that "evil is the exception."[7] By this, he means that evil only occurs here in this realm, while the large cosmos surrounding us is overwhelmingly good. While this is intellectually not difficult to understand, the implications for how this impacts human experience are remarkable. The modern experience of the universe often describes it as a cold empty space, and human existence as an accident on this "third rock from the sun." Supposedly, there is no morality to any of this: the universe is amoral, it just is. The Timaean cosmos, on the other hand, recognizes the chaos of human existence, but surrounds it in beauty and perfection. Thus, even with the evils that exist on Earth, this world is, on an overall calculus, "a happy world."[8] The evil, while real, is surrounded by good. There is as a result always an "appeal against evil."[9] Humanity can always look away from the chaos of our everyday life in the sublunary sphere, to the perfection and beauty of the cosmos.

But it is not just that humans can look to the cosmos in times of suffering and that good outweighs evil there. Brague argues that according to this model, humans are connected to and directed toward this perfection. This means both that the cosmos affects human existence without us doing anything, and also that it can further affect human existence if we strive to imitate it. That we are connected to the cosmos may seem obvious or even innocuous at first—the sun gives us seasons, the moon affects the tides, which affects when some animals breed, and so forth. The point here is that humans are also connected and influenced by the cosmos on a more profound level, more akin to astrology. Humans are a microcosm of the cosmos.[10] Just as the cosmos has a realm of Being that is ordered and perfect, we too have a soul that was often articulated as a divine or rational component of humanity. Just as the cosmos has a realm of Becoming that is chaotic, we too have a body that is changing and full of desires. As a microcosmos, our physiological structure reflects this mirroring:

> Thus in miniatures in which a man is shown nude, his arms extended to form a cross, inscribed in a square, his head surrounded by a circle, or simply inscribed within a circle—two images that a familiar drawing by Leonardo da Vinci ingeniously attempts to combine. The human form, presented as a *chi* (X), imitates not only the Cross of Christ, but the intersection of the circles of the soul which, according to the Timaeus, defines the world.[11]

Even our "erect posture" points us up toward the sky.[12] It is almost as if the cosmos is calling us: "man contains within himself what he needs to know the entire universe."[13] There is a cosmic conspiracy to help us: "Through influences, the world calls on man, as if magnetically, to allow himself to be infused by it."[14] One can see how Christians, Jews, and Muslims saw in Plato's wonderful cosmos the creation of God. It gives humans a central importance and a dignity that the modern amoral universe cannot offer. The Timaean cosmos was a home for humans.

But humans must do their part and this involves studying and imitating the cosmos. The cosmos is, on this model, "a source of morality."[15] The claim here is that the cosmos itself is good and that humans can improve their lives by imitating it. But what does it mean to imitate the stars? This question makes little sense to moderns, but in the context of the Timaean cosmos, the realm of Being is a realm of order, patterns, and regularity. The human who imitates this will seek to move away from the fluctuating desires and capricious whims of undisciplined human existence toward an orderly, regulated life like the cosmos. The point here is not just that disciplined study is valuable to becoming a disciplined person (although this may be true), but that the cosmos *itself* is harmonious and perfect, and thus imitating it can awaken an orderly life.[16]

The primary form of thinking that can accomplish this is contemplation. Here, we find yet another radical difference with modernity, which tends to treat contemplation derisively as an aesthetic and unproductive activity. We can see this in modern education, where the emphasis is on learning information and solving problems. Contemplation is an action that is fine in an art museum or a church, but irrelevant outside individual pursuits. In the Timaean cosmos, on the other hand, contemplation is not only the way to connect with the cosmos, but one of the highest forms of human activity. To contemplate the cosmos is to plug into perfection, to bask in the glory of the beauty of the cosmos—how could it not rub off on you a little? It is love—love of the world.

I want to really emphasize this point; however, there is an aspect of this discussion that needs elaboration, because, from a modern perspective, the degree and forms of connections between the individual and the world (or cosmos) being discussed here are different from what we moderns are accustomed to. Things like astrology, the idea that the stars affect our fortune, tend to be ridiculed as superstitious nonsense now, but something like astrology is being taken for granted in the Timaean cosmos. The problem here is not a debate about the reality of a human-cosmos connection, but trying to do justice to the experience of contemplating and loving the cosmos. Like colonial anthropologists, we cannot come close to comprehending what is being described here if we cannot get over, to some degree, our modern understanding. So I want to take a moment to try and make this experience more accessible.

POROUS VERSUS BUFFERED SELF

The distinction made by Charles Taylor between the "porous self" and the "buffered self" sheds more light on what it means to contemplate and love the cosmos.[17] What makes Charles Taylor unique is not only his inspiring breadth of knowledge but his interest in showing how complicated philosophical concepts are tied to issues of identity. In other words, Taylor attempts to show how the history of philosophical debates and dilemmas inevitably shapes, but is also driven by, human self-understanding.[18] The distinction between the "porous self" and the "buffered self" is probably *the* pivotal example of this for Taylor.

The "porous self" refers to the premodern self and the "buffered self" to the modern self. The distinction is between a self that is open to the world (has pores that allow from movement between the self and the world) and a self that is closed to the world (buffered or shielded from the world). This difference mirrors the change Weber famously described by contrasting the modern "disenchanted" world with the premodern "enchanted" world. According to Weber, modern science and modern bureaucracy have stripped the world of much of the magical and spiritual aspects that permeated it in premodern times.[19] Taylor is showing how this change played out with regard to the modern conception of the self and modern self-identity.

From this side of the shift, from within the buffered identity, it is difficult to fully comprehend what it would be like to have a porous self. Simply put, the porous self was open to influence from an enchanted world. This was a world of angels, demons, faeries, djinns, magic, hexes, spirits, goblins, and so forth. What are now often dismissed as superstitions—a black cat crossing one's path, the ill omen of a crow cawing, the number 6 being unlucky—in the enchanted world would have been existential realities full of meaning, both as threats and opportunities.

The premodern was much more alive in this sense. Even images (like painting or drawing or markings) could threaten or save a human. I was reminded of this recently when I observed a middle-aged woman reverently and timorously approach a painting of Jesus. She crossed herself before the image and then quietly spoke to her daughter, before lifting her daughter up to kiss the feet of Jesus in the image. This was remarkable because it showed this woman connected the image of Jesus with Jesus. Kissing the image wasn't about kissing a good work of art, it was about kissing Jesus' feet. Freedberg's *The Power of Images* shows how the premodern world took images to be directly connected to the self in this way. He gives many examples such as the Italian priests who would hold religious images, *tavlotta*, before the eyes of executed criminals because if their vision and mind was full of such images at their death they would be better off in the next life.[20] Or the disfigured aristocrat who would have his spouse look at an image of a beautiful man while they had sex in hopes that a potential child would look like the image and not its father. This was a world where images had immense power, and it is fascinating to imagine how alive and how meaningful such a world would have been. Freedberg recognizes that at first glance, this will seem quite foreign to contemporary readers, but he reminds us that pornographic or erotic images elicit powerful reactions today, so we do, on some level, have this experience of being overpowered by or resisting the power of images.[21]

Taylor's point is that as the world became disenchanted, there was a corresponding buffering of the self. This was both a product of this shift and a motivating factor in favor of it. In other words, Taylor claims that the buffered self was attractive.[22] Early modern thinkers wanted the buffered self: being cut off from ghosts and demons could seem a relief (although, it also meant losing the angels and gods). The buffered self is not just safe from the wild creatures of fantasy, but serves as an ideal of control, autonomy, order, and masculinity.[23] In other words, the buffered self was what one wanted to be "the buffered identity, capable of disciplined control and benevolence, generated its own sense of dignity and power, its own inner satisfactions."[24]

As Arendt points out, modern thinkers just want to cut themselves off not only from this superstitious enchanted world but also, in a sense, from the rest of humanity.[25] The premodern self was also connected to other human beings and therefore responsible for them and open to being hurt by them. I don't mean by this merely that another person could injure or rob them, but that the actions of a particularly inattentive or irresponsible person could anger the gods or the world in a way that had repercussions for all of the community and not just that person. Some fool offends the gods and next thing you know the town is hit by an earthquake. The buffered self, on the other hand, makes me responsible only for myself. If you anger the gods, that is on you. Arendt reminds us that premodern communities considered humans to be connected in a way that implied responsibility for each other in the sense of care, as in the Biblical admonition to take care of the widow and the orphan. But the modern buffered self, especially in its bourgeois version, allowed for some to shelter themselves from the demand to care for others. "I earned this money, it is mine, I don't owe anyone anything"—becomes a way to circumvent traditional imperatives against unfairly taking from and refusing to share with others.

The porous self is thus connected and related to both the enchanted world with all the images, fantastic beings and gods it entails and a complex, demanding, interconnected human community. Living in such an intimate, rich, spiritual existence means responsibility to and for this world, with all the monsters, treasures, and village fools that come along with it. I will not attempt to trace the complicated shifts that allowed for the rise of the modern buffered self and the disenchanted world here,[26] but I do want to tie this back to the question of how the modern buffered individual living in a disenchanted universe can better understand premodern experience. If the modern individual is not philosophically aware of these issues, their understanding of premodern life will be inevitably warped.

In fact, the buffered self is itself a delusional myth. This is one of the repeated central claims of Taylor's works: there is no such thing as an autonomous, independent, "free" individual.[27] Upholding this ideal and incorporating it into our social and political theories has had devastating effects. Taylor's point is that despite compelling arguments against the reality of the autonomous individual, the ideal and prestige of the autonomous individual has impeded a full recognition of its problems.[28] This is a problem for political and social theory but also for recognizing and experiencing *amor mundi*. I will return to this in chapter 6.

The "porous self" and the "buffered self" are, like most distinctions, probably best viewed as a spectrum, where the modern self has moved toward the latter end. For most of us raised in the context of rationalization, science, capitalist ideologies of "autonomous" individualism, and disenchantment, it is difficult to comprehend the experience of the porous self. It is not entirely foreign however—as Freedberg pointed out, imagery can still arouse us and produce powerful effects.

With the "porosity" of premodernity in mind, let us return to the issue of the contemplation and love of the cosmos.

THE TIMAEAN COSMOS: CONTEMPLATING THE COSMOS

In *The Human Condition*, Arendt argues that from Plato to modernity the West was dominated by a period of *vita contemplativa* as opposed to the democratic *vita activa* of pre-Platonic Greece and the science, technology, and capitalist *vita activa* of modernity.[29] In other words, from Plato to modernity, the supreme philosophical act was one of contemplation. Now, we tend to think of contemplation as being lost in thought or speculation, but Brague shows that for prominent figures like Seneca and Plotinus "to be concerned with the objects of contemplation is to be actively involved in the highest form of politics."[30] Contemplation on this model is the highest form of action (praxis). This should appear odd to us now, as we tend to separate theory from practice and treat theory as a means to improved practice. In other words, we tend to think of theory as empty unless it has practical implications. But Seneca is claiming that contemplation *is* a form of action. What this means is that contemplation is an absolutely fundamental part of a complete life: "I live according if I give myself to it entirely, if I devote my life to admiring and adoring it."[31] If you want to make life better (for yourself and for all of us), contemplate the cosmos.

Contemplation involves opening oneself up to the cosmos. For Christians, Muslims, Jews, and Stoics, the cosmos was connected to God—either as a creation of God that reflected His goodness or as "indistinguishable from God" in the case of later Stoics.[32] "The dignity of contemplation and the dignity of the world are mutually reinforced: it is because contemplation is the loftiest activity that it has bearing on the world, reciprocally, it is the supreme dignity of the world that endows contemplation with all its worth."[33]

The experience of contemplation was a moment of communion and connection with the cosmos. Done well, it is an act of love. The individual, porous and open, could be improved by basking in its glory. This is also a way of life, and one could structure one's life by imitating the cosmos. As we saw earlier, a human was a microcosmos that mirrors the macrocosmos. By striving to put their "conduct in order in all ways" they could "make themselves a little world."[34] This odd phrase shows how love of the cosmos can help an individual become a little cosmos. This is both an individual and a communal task. Brague points out that this demand was also applied to cathedrals and city planning.[35] Ancient thinkers asked themselves, "What would the cosmos do?"

Imitating the cosmos was a matter of study. "The world is full of meaning, and of a meaning that man is capable of deciphering and applying to himself. Wisdom would mean a wisdom of the world in that it would consist in correctly interpreting the messages contained in things."[36] For the Christian, Muslim, or Jew, the world is one of the ways God "presents us with models of how we should behave."[37] Since the cosmos is thoroughly moral, "the moral virtues are present in things."[38] Through study, practice, and loving contemplation of the cosmos, we can achieve the virtue it embodies.

My approach here has been abstract, philosophical, and academic. This is, I suspect, inadequate. It is one thing to be able to intellectually articulate an alternative metaphysical system and its implications for epistemology, ethics, and human nature. But, as anthropologists, sociologists (e.g., Bourdieu) and phenomenologists have pointed out, this will always fall far short of the lived reality.[39] As such it risks being superficial and failing to do justice to phenomena in question. Media like literature and film can connect a viewer more viscerally than standard philosophical analysis.

A literary, film-based, and/or comparative anthropology-style analysis goes beyond the scope of my purposes here. What I have attempted to do is clarify a certain kind of *amor mundi*—love of the cosmos—that Brague claims was standard in western intellectual thought from Plato to Copernicus. What I have attempted to show so far is the metaphysical structure of the Timaean cosmos and some of the implications for ethics and human nature. My focus has been on the *mundi* part of this kind of *amor mundi*. In what remains of this chapter, I want to focus on the *amor* part—on love.

Like the discussion of the ancient cosmos, my purpose is partially historical. I will present a historical conflict about love in Plato. But in so doing, I will also attempt to open up questions that are fundamentally important now. By seeing the struggle in Plato concerning love, we can then think about how we conceptualize love now and what it would mean to take the task of *amor mundi* seriously.

LOVE AND KNOWLEDGE: GNOSTIC ANTI-COSMISM AND ANCIENT DOUBTS ABOUT SEXUAL LOVE AND EMOTION

We have just worked through Rémi Brague's description of the model of the cosmos presented by Plato in the Timaeus, which he claims became one of the primary intellectual understandings of the cosmos in the West from Plato until Copernicus. We have also seen some of the implications of this model for ethics and human nature. My claim is that imitating and contemplating the cosmos is a kind of *amor mundi*—love of the cosmos. This language of love is not used by Brague however. Perhaps the term "love" is not used because of Plato's ambivalence and early antagonism toward love—an ambivalence that was widespread in the ancient Western thought.[40] This is unfortunate because Plato is worried about sexual love and passion while seeming to nonetheless suggest a kind of loving contemplation or contemplative love. Thus the remainder of this chapter concerns what it means to *love* the world. I will begin with the negative evaluation of sexual love and the seeming antagonism between love and knowledge in Plato and other ancient thinkers.

As I have mentioned, Brague argues that the Timaean model is only one of the ancient models of the cosmos. Of the models he presents—Timaean, Atomist, Scriptural, and Gnostic—it is the Timaean that most demands *amor mundi*. Interestingly the Gnostic model serves almost as an inversion of the Timaean model. While there are a variety of Gnosticisms,[41] Brague claims the basic outlook in Gnosticism is that this life a trap, that humans have a spiritual essence, a soul, that has fallen from its previous spiritual state and become encumbered with a material body and material desires.[42] As Hans Jonas points out, this was sometimes conceived as a trap made by the God of the Old Testament, Jehova.[43] Jesus, against this malicious God Jehova, has come to let humans know the truth of their situation: thus the name *Gnosticism*, which refers to the secret knowledge of this predicament. Knowing that this life is actually a trap, a kind of hell, is, of course, the necessary first step to escaping this conundrum.

Like Plato, Gnosticism takes the human body and human desires to be a cause of this chaos, but in addition sees it as that which traps us here. We are caught up and distracted by our physical desires in this delightsome hell. The basic model of the Earth-centered cosmos made of concentric spheres remains, but the goal is not merely to imitate and contemplate the cosmos, but to escape the Earth and return to more spiritual realms. Gnosticism seems motivated not by love of the cosmos, but by hatred of the Earth, our bodies, and our physical desires. The tenor of Gnosticism is despair, secrecy, and negation (which prompts Jonas, rightly or wrongly, to compare it to modern existentialism).[44]

Note that Plato shares with Gnosticism a suspicion of the physical body, physical desires, and emotions. This creates a tension, as emotion and physicality need to be corrected or controlled by imitation of the cosmos, and yet contemplation of the cosmos is itself an ecstatic experience—a love that is one of the most sublime experiences possible for humans. Due to this ambivalence, Plato vacillates between praising and condemning love at different points. This tension regarding love and emotion is what I will be exploring in the remainder of this chapter.

This suspicion of emotion was also shared by the Stoics. According to Furtak:

> Emotions color human experience in various ways: they seem to be connected with whatever matters most to us, and yet they are often so unpleasant or disconcerting that we find ourselves wishing we could be rid of them altogether. According to Stoic moral psychology, emotions (or passions) are cognitive responses to perceived value in the world, and therefore we can eradicate them by changing our beliefs about that really matters in life. Stoicism is the ancient source for a perennial bias against the emotions, and it has so thoroughly lefts its mark on conventional wisdom that it is acceptable in contemporary English to use the word 'philosophical' to describe a person with a stoical or calm disposition.[45]

As Furtak argues, this suspicion of emotion remains in Western thought, and we don't have to look hard to find it. Open up a high school science textbook and look for a discussion of love. It doesn't occur. Love apparently isn't relevant to science. Love is often experienced and described as an emotion—a particularly strong emotion. Science, as it is popularly described, is supposed to be an objective, verifiable form of obtaining knowledge. As such, we are taught, love and other strong emotions need to be carefully controlled to preserve the "objectivity" and integrity of scientific experimentation.

Postpositivist philosophy of science has shown this image of unbiased, objective science is not true to actual scientific practice.[46] Science, it turns out, is not an emotionless, purely objective enterprise—good science requires judgment, emotion, and care.[47] Nevertheless, the ideal and image of emotionless, rational objectivity remains a force in western society and thought. Feminist philosophy in particular has done much to critique this supposed opposition between emotion and rationality. This is not surprising as the distinction is often gendered with women associated with "inferior" emotion and men with "superior" rationality and objectivity.

I want to dwell with these tensions about love. While there are a variety of sources we could use to think about love, I want to keep the topic close to Plato since we have been discussing his model of the cosmos. This is not new philosophical territory and I will rely on Martha Nussbaum's *The Fragility of Goodness: Luck and Ethics in Greek Tragedy and Philosophy*. Nussbaum is an important figure in calling philosophy out for its lack of attention to love and literature. As can be seen by the title of one of her earlier works, *Love's Knowledge: Essays on Philosophy and Literature*, Nussbaum has argued for the importance and intelligence of love and emotion and the philosophical relevance of literature.[48] Instead of the traditional philosophical distancing of love and knowledge, Nussbaum argues that emotion and literature are important kinds or aspects of deep philosophical thinking.

In *The Fragility of Goodness*, she offers a somewhat controversial reading of Plato that shows the tension in Plato's own attitudes toward love and emotion. Nussbaum argues that in early and middle Plato we find an antagonistic attitude toward love and emotion and a more ambivalent attitude toward love in the later works. The Timaeus falls closer to the later works and the ambivalence of later Plato than the attitude of early Plato. The narrative Nussbaum presents is one of Plato taking a strong stance against emotion, only to later realize this was not correct. While some might find this too coherent a narrative, I think that it nonetheless shows the tensions surrounding Plato's conception of love.

Let me turn to her analysis.

EARLY PLATO AGAINST LOVE

Nussbaum argues that ancient Greek society and by extension Greek art was concerned with luck, *tuchē*, or the randomness of life. Shit happens. Against this rose "the aspiration to rational self-sufficiency in Greek ethical thought: the aspiration to make the goodness of a good human life safe from luck through the controlling power of reason."[49] The key to controlling *tuchē* are *technai*, human arts and/or sciences. Here "art" refers to the creation of tools, crafts, and useful objects that help us control our surroundings. Building a house protects against the weather, building a wall protects against enemy armies, weapons can protect against wild animals, and so on. Similarly appropriate knowledge can help humans comprehend and deal with life difficulties appropriately.

Nussbaum claims that Athens at the time Plato was young was full of "acute anxiety and of exuberant confidence in human power."[50] Athens was excited by the possibility of progress and overcoming contingency but also aware that this could go awry. Famous tragedies like *Antigone* "warned against overambitious attempts to eliminate luck from human life."[51] Nussbaum claims that Sophocles contrasts a desire for simplification, order, and control with an "open responsiveness" that works with others and the challenges of life.[52] Creon in the play represents an active "rage for control" that becomes rigid, narrow-minded, and obsessive. If we are not careful, our vision of order can be so inflexible that we attempt to force it on reality and on others in a way that is unresponsive and destructive. On the other hand, there is a more passive recognition of the complicated world around us that seeks not to tyrannically impose its will on life, but learn, flexibly work with and carefully respond to life's challenges. The risk here is being too passive, letting oneself be swept away with the stream. Nussbaum's claim is that *Antigone* argues for a balance between "self-protection" and "yielding" because "the completely passive victim cannot act to help another" while the "Creonic agent cannot see otherness."[53] Thus, there is a need for a middle way, a "delicate balance between order and disorder, control and vulnerability."[54]

Plato, in this account, rejected this message. In his early and middle writings, he becomes a key proponent of *technē* as a way to improve the human condition. But Plato suggests a specific kind of *technē*, "one that assimilates practical deliberation to counting, weighing, and measuring."[55] In other words, according to Nussbaum, Plato wants to use quantitative measures to understand the world and resolve conflicts.

For Plato, the problems in this world of constant change come from within and without: human passion and human desire for food, sleep, and sex can lead to disorder and chaos, in addition to the threats from weather, barbarians, disease, and so on. *Technē* can provide security against the outside world, but it must also modify the self. Like Alcibiades of the *Symposium*, if we allow ourselves to be ruled by desire, in this case his erotic love for Socrates, we risk being completely disorganized and out of control.[56] This can affect both the individual and an entire society. We have already seen that in Plato's Timaean cosmos, the entire sublunary realm is the realm of Becoming and change. Gnostics simply attempted to leave this realm, while Stoicism sought to gain control at an individual level. In the *Republic*, which predates the *Timaeus*, Plato famously describes how to create an ordered society with the goal that it be ruled by reason and therefore ordered and perfect.[57] This requires overcoming the appetites and emotion that too often rule human life instead.

The position here is opposed to love, especially erotic love, as an emblematic destabilizing force. Things like "eating and sexual activity are unstable for two reasons: because of their internal structure and because of the nature of their objects."[58] Not only can erotic love itself be unruly, unpredictable and wild, but the object of love may not reciprocate and is a particular being that will die. If we want stability, we need to aim for something eternal and have a method to approach or obtain it that is sure. At this point, Plato's theory of Forms and the intellectual soul come into play, but this goes beyond what we need to understand here. To be brief, Plato claims intellectual truth is eternal and reason is the method to obtain it. This should be the aim of any search for stability and happiness.

For Plato, the philosopher that seeks truth cannot be distracted by hunger and sexual desire, instead they must get themselves "to a point at which [they] have no pressing needs at all and can therefore survey all the alternative activities coolly, clearly, without pain or distraction, 'using pure reason itself by itself' (Phd. 66A)—so that eating and love-making do eventually look, to [their] soul's eye, like nothing more interesting than the grazing and copulating of cattle."[59] The philosopher becomes like the eternal objects they contemplate: "pure, hard, single, unchanging, unchangeable."[60] No longer distracted by material desires, the philosopher is now "ruled by reason" and in a position to objectively evaluate, rank and "order alternative pursuits."[61]

This would seem to make philosophers the ideal rulers for Plato's ideal city. After all, Plato thinks, "a human who is weaned from all attachment to internally unstable pursuits such as love, sexual activity, power-seeking, and money-making is automatically at the same time rid of many of the most common grounds of value conflict."[62] The philosopher can supposedly see the different values at play in a particular conflict more clearly, since they have reached a state of distance from mundane concerns. This would make them good judges and rulers. The problem is that philosophers are more interested in contemplating eternal truths than mediating mundane conflicts. The philosopher has learned to get over petty issues: "In part, the superior harmony of the philosopher's life results directly from this reduction in the number of his or her commitments."[63] Thus Plato claims that while philosophers would be the best rulers, they are not going to be interested in doing so. Contemplating eternal truth is much more interesting than fixing silly problems one has already risen above.

But how does this rational accounting resolve conflicts? Through quantification and math. Nussbaum argues that in the *Protagoras* Plato presents a new *technē*—a way to solve practical conflicts through "counting, weighing and measuring."[64] In the *Protagoras* Plato has Protagoras and Socrates present two different models for a "*technē* governing practical choice."[65] According to Plato's Socrates, in order to have a "systemic grasp" on the world around and "master contingency" we need a *technē* that is universal, teachable, precise, and concerned with explanation.[66] Sounding like a proto-economist or proto-utilitarian, Socrates argues that an "ethical science of measurement" can find a way to "render values commensurable."[67] In fact, Nussbaum argues that the approach here is quite similar to that voiced by utilitarians like Bentham and Sidgwick. She quotes Sidgwick, who complains against detractors of utilitarianism that unless we find something like happiness that can be used as a "common end" that is shared by all people, we will never be able to "systematize human activities" in a way that allows for ethics to be "scientifically complete."[68]

Is it appropriate to read Plato alongside utilitarians? Some scholars might worry there is a risk of reading utilitarian concerns back into Plato. Correct or not, Nussbaum thinks Plato is grappling with similar concerns and suggesting a similar response: against the potential chaos of multiple value commitments, we need common measures that can allow for comparison between them. Utilitarianism, of course, suggests that equations of happiness or pleasure and pain are the best option to do so. Bentham argues for a hedonistic calculus that would enable us to calculate out the rational best option. If I want toppings A, B, and C on my pizza and you want toppings C, D, and E, but hate topping A, we can affix numbers that reflect how much we like or dislike each topping and calculate out a pizza that maximizes the amount of happiness we can share. Economists use money in a similar fashion to measure how much people desire certain outcomes. Money certainly has a disturbing ability to make things commensurable that might seem incommensurable at first. And this is the power of such models: they reduce hypercomplex issues into commensurate, quantified systems. Through math and science we can supposedly "change our world" to make it more rational, more transparent and more capable of resolving conflict.[69]

In the *Protagoras*, Plato's Socrates never offers a complete system for adjudicating different values along the lines of utilitarianism or modern economics. On Nussbaum's reading, he argues for the importance of such a system without offering the specific details. What he does do is connect the potential of such a system to the "transformation of our lives."[70] Like utilitarianism, this transformation not only occurs at the level of relationships between people (helping mediate conflicts) but also transforms individuals themselves. In line with the concerns of the *Republic*, "the acceptance of qualitative singleness and homogeneity of the values actually modifies the passions, removing the motivations we now have for certain sorts of irrational behavior."[71] In moments of conflict, the philosopher that is armed with this kind of *technē* can calmly and rationally work through the (quantified) options and calculate the best results. Mathematics then, is a kind of cooling force. It tempers the emotions and helps us become rational. Plato thinks that there is a part of all humans that is drawn to "mathematical, scientific and philosophical reasoning" because it is "enormously beautiful and compelling to human souls."[72] Tradition says that Plato's Academy has the inscription "Let no one enter here who does not pursue geometry"—which might seem strange unless one understands the transformative power of math and the immense "joy and pleasure" of "pure reasoning."[73] Such an inscription is an invitation to happiness.

LATER PLATO'S AMBIGUITY ABOUT LOVE

This language of "compelling," "enormous beauty" sounds like the attraction and desire of erotic love. Ostensibly, however, we are being turned away from wild erotic love of a particular, to a calm, contemplative love of "pure reasoning" (and later the cosmos as a seeming embodiment of reason and order). It would seem that we are dealing with two different kinds of love, an irrational erotic love that is denigrated and a rational contemplative love that is extolled. We are turned away from one toward the other. But there is something strange here, because Plato is using the intense language of erotic love and directing it toward a calm, contemplative love. Nussbaum points out that Plato also uses the tools of the literary tradition he is criticizing to garner enthusiasm for his anti-literary positions.[74] The artists must be kicked out of the ideal city because of their power to create illusions and their ability to sway people to bad ideas, and yet Plato himself writes with compelling literary force.

As a result of these tensions, it is initially confusing when Socrates says in the *Symposium* that "The only thing I say I know is the art of love."[75] Has Plato not been telling us how bad love is? And now he has Socrates say this is the only thing he knows? Of course, we discover in the *Symposium* and the *Lysis* that what Socrates means by this is a "hunger for wisdom." Hunger and desire, which we are first taught to learn to quash, are now reawakened but aimed at wisdom and truth instead of food and sex.

Nussbaum finds this denigration of the "animal" appetites (including erotic love), emotions and the physical body in favor of "pure" intellectual, nonmaterial elements to be woefully inadequate. She is especially frustrated with how this functions as a quest for "self-possession."[76] Socrates, in rejecting appetite, body, and emotion and ascending toward the rational and intellectual "has become, himself, very like a [Platonic] form—hard, indivisible, unchanging . . . he is stone; and he also turns others into stone."[77] The famous legends about Socrates being impervious to cold are not a sign of triumph, but a sign of something not "fully human" about Plato's philosophy.[78] "Inside the funny, fat, snub-nosed shell, the soul, self-absorbed, pursues its self-sufficient contemplation."[79]

Against Plato, Nussbaum argues that the erotic lover is not the fool he is presented to be. There are "particular truths and particular judgements" that cannot be learned by "Socratic adherence to rule" and fixation with universal principles or forms.[80] Alcibiades, who is out of control and mad with love for Socrates in the *Symposium*, is presented by Nussbaum as instead offering "a kind of practical understanding that consists in the keen responsiveness of intellect, imagination and feeling to the particulars of a situation."[81] Emotions, body, and intellect can and should work together.

Nussbaum claims that Plato realizes this to some degree in his later work and she sees the *Phaedrus* as a key moment of recantation. This can be seen in the structure of the Phaedrus which offers two speeches that are opposed to the mania of love followed up by Socrates recanting his anti-love speech with another that argues in favor of the mania of love. The dialogue finds young Phaedrus walking with Socrates outside the city. Phaedrus has just heard a speech by Lysias that has impressed him and Socrates asks him to repeat it (note that Socrates says he is "sick with passion for hearing speeches"). Socrates then offers his own speech. Both offer advice to Phaedrus. Nussbaum summarizes both speeches as follows: "cultivate in yourself the state of self-possession, *sōphosunē*. Develop the clarity of your intellect by exercising strict control over bestial non-intellectual elements. Form only non-mad friendships, and only with other self-possessed, non-mad people."[82] On her interpretation, this is more of the same stuff Plato has been preaching.

What is surprising is that Socrates suddenly stops and claims he has made a mistake. Having bad-mouthed love, his *daimon* (spirit guide) says he needs to atone to *Eros* (242c). Socrates offers a new speech and this time he praises erotic mania. As long as they are divinely inspired, Socrates claims there are multiple forms of madness that are good for humans (244a-245c) and praises erotic madness as the highest form of divine madness (249d-e). He also offers his famous chariots analogy of the soul, in which he compares the soul to a chariot pulled by two horses. One horse is "vicious," "heavy," and "weighs the charioteer down" (247b). The other is noble and less wild. The traditional interpretation of this allegory is that the charioteer is the rational intellect which is pulled along by the wild horse, which represents ignoble emotions, the material body, and material desires for sex and food, and the noble horse which represents noble emotions and passions. While the physical body is still treated as the inferior horse, Plato nonetheless puts it in a position to propel the soul forward.

Nussbaum reads this recantation in the *Phaedrus* as a qualification of Plato's previous obsession with rational self-possession and denigration of love. While still requiring it be controlled by the rational intellect, Plato seems to be saying that there is a place for divine erotic madness. The question is no longer one of rejecting love and quashing material desires, but of directing their energies. Instead of a quest for stone-like self-possession, Plato is open to receptivity to others, to erotic madness for others: "the truly blessed life involves the proper cultivation of both activity and passivity, working in harmony and mutuality."[83]

This is a rather abrupt and quite surprising shift. As a translator of the *Phaedrus* Stephen Scully comments, "the tone and language in the palinode [the second, pro-love speech] . . . exceeds anything found in the rest of Plato."[84] It seems to come out of nowhere. For Nussbaum it represents an important moment of recantation by Plato, or at least ambivalence.

With regard to love, Irving Singer in his three-volume history of love in Western philosophy confirms this ambivalence:

> Analyzing sexual love in his own way, Plato treats it with his characteristic ambiguity. At times, he sounds as if sex and love were largely incompatible: since love is directed towards absolute beauty beyond the empirical world, the enlightened philosopher must renounce physical interests that tie him to mundane sense experience. When self-indulgent Alcibiades depicts the sexual continence of Socrates and remarks that between them there can never be peace, Plato seems to be saying that the very presence of carnal appetite interferes with the philosophical attitude in which true love consists. At other times, however, Socrates is shown to have erotic interests in young men. In several places we are told that only a god could live a life of pure contemplation: since the philosopher is human like everyone else, he cannot eradicate his material nature but must satisfy it while possibly conforming to the spiritual principles in which he believes. A similar ambiguity arises in medieval thought. Some theologians in the Middle Ages considered sexual impulse sinful, irreconcilable with the love of God even when directed towards one's spouse. Other theologians, Aquinas for instance, maintained that desire becomes evil only if it subdues man's higher faculty of reason. Dante argued that, once it has been rendered subservient, sexual inclination may even further the love of God. But though Aquinas and Dante recognize the need to harmonize sexual and religious interests, they reinstate the Platonic ambiguity by portraying the former as forever problematic and certainly unworthy of being idealized in itself.[85]

As I mentioned earlier, Nussbaum's early work, including *The Fragility of Goodness*, is focused on the importance of love and emotion for knowledge. She seeks to push back against a philosophical tradition that looks to science (especially physics) as the outstanding academic discipline. Meanwhile, literature and the humanities are considered fluffy, irrelevant, and unscientific. In a sense, Nussbaum is attempting to fight the powerful, continuing influence of the anti-love, anti-art Plato with the divine erotic madness that is extolled in the *Phaedrus*. Against turning ourselves into self-possessed stones, Nussbaum calls for openness, relationality, and love. And when we do turn outward toward the world, instead of forcing it into the order dictated by intellect, we should responsibly and flexibly work with it. And instead of a cold, rational, distant, contemplative love, she praises the energetic passion of erotic madness.

I want to stress that what sets her reading apart is Nussbaum's insistence on love as erotic. It is too easy to read what she is saying and miss this—after all, we teach our students that philosophy means love of wisdom, so of course love is involved. In my own initial experience reading this text, I thought, yes, philosophy is a kind of love. But without realizing it, I was taking for granted that this means something like the cold, rational, contemplative love that Plato has for intellectual things. After all, philosophy is called *philo-sophia*, not *eros-sophia*. There is an issue of performativity here also. Interestingly, while Plato skillfully uses the literary arts against the poets to extoll cold, rational philosophy, Nussbaum writes in the cold rational style of science-loving philosophy to extoll the need for love, emotion, and literature. She is working from inside a tradition that seeks to be rational and objective, using the coldly rational methods of this tradition, to attempt to get it to see the flaws of this kind of approach. I am not particularly committed to cold rationality, objectivity and science. I love literature. And yet I was so used to this style and these values functioning tacitly in philosophy, that it was a moment of shock when I finally got it: seeking wisdom should be like an erotic madness. It should overwhelm and suffuse us.

LOVE OF THE COSMOS

The term "*amor mundi*" comes from Arendt. As we will see in chapters 3 and 4, she uses it to describe love for a particular sociopolitical and cultural world. The term is not used by Plato, Brague or Nussbaum. Is it an appropriate term to apply to the imitation, contemplation and adoration of the cosmos that Brague describes as the "standard vision of [and mode of comportment toward] the world" in the premodern West?

Despite Plato's reservations about love, I think this is clearly a correct way to describe how one should relate to the cosmos if one accepts the Timaean model. And while the Western philosophical tradition does manifest a deep ambivalence about love, we find thinkers that are caught up in erotic madness alongside the more stoic, contemplative lovers of the world. What both of these forms of love have in common is connection and relationship. For Medieval Christians, either form of love was often connected with God, but of course for the porous premoderns living in an enchanted cosmos, God was everywhere—the cosmos was His creation and its goodness pointed to him. St. Francis of Assisi speaks to birds. Meister Eckhart and other "mystical" thinkers stress ecstatic experience with God and the world. Other thinkers express suspicion of this kind of madness.

Whether carried away or lovingly contemplative, there is definitively a kind of love of the cosmos at work here. And it is not something merely interesting, but a fundamental philosophical act. Arendt characterizes the entire epoch between Plato and modernity, almost two thousand years, as the time of the *vita contemplativa* wherein the highest act was contemplation. As we have seen such contemplation is a kind of *amor mundi*—a dwelling with, contemplation of and joy in the cosmos.

But if it is the case that the highest act for two thousand years of Western thought was loving the cosmos, what happened to it? What has happened to this supposedly fundamental experience? In the next chapter, I will attempt to show what led to the decline of cosmic *amor mundi*.

NOTES

1. Brague, Rémi. *The Wisdom of the World: The Experience of the Universe in Western Thought* (Chicago: University of Chicago Press, 2003).
2. Brague, *The Wisdom of the World*, 85.
3. Ibid., 88.
4. Arendt, Hannah. *The Human Condition* (Chicago: University of Chicago Press, 1998).
5. Brague, *The Wisdom of the World*, 88.
6. Ibid., 111.
7. Ibid., 108.
8. Ibid., 106.
9. Ibid., 113.
10. Ibid., 98.
11. Ibid., 94.
12. Ibid., 99.
13. Ibid., 95.
14. Ibid., 98.
15. Ibid., 115.
16. Ibid., 38, 115.
17. Taylor, Charles. *A Secular Age* (Cambridge, MA: Harvard University Press, 2007).
18. Taylor, Charles. *Sources of the Self: The Making of Modern Identity* (Boston, MA: Harvard University Press, 1992).
19. Weber, Max. *Economy and Society* (Berkeley, CA: University of California Press, 1978).
20. Freedberg, David. *The Power of Images: Studies in the History and Theory of Response* (Chicago: University of Chicago Press, 1991), 6.
21. Freedberg, *The Power of Images*, 2.
22. Taylor, *A Secular Age*.
23. Lloyd, Genevieve. "The Man of Reason." *Metaphilosophy*, Vol. 10, No. 1 (January 1979).
24. Taylor, *A Secular Age*, 262.
25. See Chapter Two in Arendt, *The Human Condition*.
26. See Taylor, *A Secular Age*.
27. Taylor, Charles. *Philosophical Papers: Volume 2, Philosophy and the Human Sciences* (New York, NY: Cambridge University Press, 1985).
28. Taylor, *A Secular Age*.
29. Arendt, *The Human Condition*.

30. Brague, *The Wisdom of the World*, 124.
31. Ibid., 125.
32. Ibid., 122.
33. Ibid., 122.
34. Ibid., 133.
35. Ibid., 135.
36. Ibid., 119.
37. Ibid., 119.
38. Ibid., 119.
39. See, for example, Bourdieu, Pierre. *Practical Reason: On the Theory of Action* (Stanford, CA: Stanford University Press, 1998) and Heidegger, Martin. *Being and Time* (London: SCM Press, 1962).
40. Singer, Irving. *The Nature of Love: The Modern World* (Boston, The MIT Press, 2009), 9.
41. Jonas, Hans. *The Gnostic Religion: The Message of the Alien God and the Beginnings of Christianity* (Boston: Beacon Press, 2001).
42. Brague, *The Wisdom of the World.*
43. Jonas, *The Gnostic Religion.*
44. Ibid., 320.
45. Furtak, Rick Anthony. *Wisdom in Love: Kierkegaard and the Ancient Quest for Emotional Integrity* (Notre Dame: University of Notre Dame Press, 2005), xi.
46. See for example Polanyi, Michael. *Personal Knowledge: Towards A Post-Critical Philosophy* (Chicago, IL: Chicago University Press, 1974).
47. Harding, Sandra, ed., *The Feminist Standpoint Theory Reader* (New York: Routledge, 2004).
48. Nussbaum, Martha. *Love's Knowledge: Essays on Philosophy and Literature* (New York: Oxford University Press, 1992).
49. Nussbaum, Martha. *The Fragility of Goodness: Luck and Ethics in Greek Tragedy and Philosophy* (New York, NY: Cambridge University Press, 1989), 3.
50. Nussbaum, *The Fragility of Goodness*, 89.
51. Ibid., 89.
52. Ibid., 78.
53. Ibid., 81.
54. Ibid., 81.
55. Ibid., 90.
56. Plato. *The Symposium*, trans. with commentary by R. E. Allen (New Haven: Yale University Press, 1993).
57. Plato. *The Republic*, trans. with commentary by R.E. Allen (New Haven: Yale University Press, 2006).
58. Nussbaum, *The Fragility of Goodness*, 149.
59. Ibid., 155.
60. Ibid., 138.
61. Ibid., 138.
62. Ibid., 158–159.
63. Ibid., 159.
64. Ibid., 90.

65. Ibid., 94.
66. Ibid., 95–96.
67. Ibid., 109–110.
68. Ibid., 112–113.
69. Ibid., 120.
70. Ibid., 117.
71. Ibid., 115.
72. Ibid., 162.
73. Ibid., 162.
74. Ibid., 112.
75. quoted in Reeve, C. D. C., "Plato on Friendship and Eros," *The Stanford Encyclopedia of Philosophy* (Summer 2016 Edition), Edward N. Zalta (ed.), https://plato.stanford.edu/archives/sum2016/entries/plato-friendship/?/. Accessed August 2018.
76. Nussbaum, *The Fragility of Goodness*, 206.
77. Ibid., 191.
78. Ibid., 198.
79. Ibid., 183.
80. Ibid., 190.
81. Ibid., 191.
82. Ibid., 206.
83. Ibid., 251.
84. Plato. *Phaedrus* (Cambridge: Focus, 2003), 81.
85. Singer, *The Nature of Love*, 9.

Chapter 2

The Rise of Science, the Death of the Cosmos, and the Decline of *Amor Mundi*

In chapter 1, I argued that one of the highest forms of human activity in the premodern West was *amor mundi*, love of the cosmos. This particular form of *amor mundi* was based on the Platonic model of the cosmos outlined in the *Timeaus*. For almost two thousand years, from Plato to Copernicus, this model was a powerful influence on intellectual circles, and it encouraged a contemplative love and imitation of the cosmos. There were, however, tensions about love in the premodern world. Erotic love, emotions, and physical desires were often denigrated as irrational and dangerous. So Platonic and medieval *amor mundi* was generally taken to be a kind of contemplative, rational love of the cosmos. At times, it could also manifest as a divinely inspired, passionate, erotic love.

If this is correct, then the question that arises is: what happened to this cosmic *amor mundi*? We hear nothing about it now. Our science textbooks say nothing about love. The Platonic cosmos is largely forgotten. The idea of finding ethics in the sky is now absurd. What happened?

Understanding the death of the cosmos in its historical and philosophical context is fraught because the meaning of this event is disputed and different groups want to use the event for different purposes. Broadly speaking, the death of the cosmos is associated with the collapse of the medieval world and therefore related to the pivotal shift between premodernity and modernity in the West. Thinkers that are associated with the beginning of modernity such as Descartes, Bacon, and Hobbes articulated modernity as a new beginning that broke with the past. They sought to remake society with new methods of knowing and new forms of social organization. Following thinkers such as Jürgen Habermas, Santiago Castro-Gómez, and Zygmunt Bauman, I will call this radical plan to remake society the "project of modernity."

From this side of the project of modernity it is common to follow Descartes and Hobbes in taking a pejorative attitude toward both Western and non-Western premodern societies. That the Middle Ages are often called the Dark Ages is not a simple truth about the past, but an attack on it. Calling premodern societies or peoples "primitive" is a way for moderns to help themselves feel "advanced." But critics of the project of modernity would claim this covers over the injustices of modernity (calling the colonization of the Western United States "manifest destiny" instead of genocide, for example) and presents a radical break with premodernity when there is no such radical break. Naively accepting the narrative of modernity and modernization obscures more than it reveals. Some scholars have pointed out the ways that modernity has roots in medieval thought,[1] while others have rejected the tendency to associate modernity with secularism and shown that modernity was just as much a religious project as a secular one.[2]

Depending on how one views these changes, the death of the cosmos could be seen as the end of premodernity. It is clear that scientific discoveries undermined the Timaean cosmos. New technologies led to new discoveries such as the existence of moons orbiting Saturn and Jupiter, imperfections on the moon, and comets. The no longer tenable ancient model of the cosmos was replaced by a modern universe. This metaphysical shift forced changes in epistemology and left ethics undermined. The understanding of what it meant to be a human was altered and the relations of humans to the cosmos/universe also shifted. In short, the death of the cosmos has led to radical changes that we are still struggling to comprehend.

In the first part of this chapter, I will begin by discussing the rise of the project of modernity and some of the philosophical and sociopolitical implications of the narratives and language being employed to articulate this shift between premodernity and modernity. With some understanding of these tensions, I then turn to the difference between the ancient cosmos and the modern universe, both with regards to their structure and the metaphysics involved. I will then discuss the epistemological shift from love to science. Then I will discuss the implications for modern ethics. And finally, I will discuss the implications for modern understanding of human nature. As we saw in the last chapter, metaphysics, epistemology, ethics, and human nature were all connected in the Timaean cosmos. We will see that in modernity they continue to be related and influence each other. In other words, a radically different understanding of the cosmos reverberates through epistemology, ethics, and anthropology. Obviously, these are immensely complicated changes and my aim here is a general indication of some fundamental changes that are relevant to the question of *amor mundi*.

The second half of this chapter will discuss the implications of the death of the cosmos from three different perspectives: the Death of Nature

(Merchant), the Death of God (Nietzsche), and the ontological dominion of death in modern thought (Jonas). I conclude with a discussion of Nietzsche's call for *amor fati* (love of fate).

THE DEATH OF THE COSMOS AND THE PROJECT OF MODERNITY

As odd as it sounds, the death of the cosmos was related to improvements in lens crafting. The telescope and the microscope both revealed aspects of the world around us that undermined the traditional understanding of the cosmos. For example, with the aid of a telescope, it was discovered that there were moons orbiting Jupiter and Saturn, sunspots on the sun, comets that travelled through the spheres, imperfections on the moon, the appearance of new stars, and so forth.[3] This showed that the realm of supposedly unchangeable Being above the sphere of the moon was full of all sorts of change. The cosmos above was not like Plato (or Ptolemy or Aristotle) described. Furthermore, the microscope opened up a whole microscopic world that was full of all sorts of strange things. Plato had simply been wrong.

The death of the cosmos is often associated with "truths" "discovered" thanks to the scientific revolution and new technologies. I put "truth" and "discover" in quotes here not to deny that it is true the Earth orbits the sun nor that this was discovered through careful gathering of data, but to indicate that how the many different aspects and features of the scientific revolution are articulated, marshaled, and narrated is highly controversial, both philosophically and politically. Much is at stake in how these shifts are understood and taught.

For example, perhaps the most common understanding/articulation of "what happened" with the death of the cosmos and its replacement by the universe is that new information was discovered thanks to new technology that "revealed" the truth that had been previously misunderstood. This is how scientific discovery is almost always articulated in public school textbooks, popular understanding, and even perhaps most scientists themselves. The basic narrative arc is that there was a state of ignorance or misunderstanding and through scientific experimentation the truth was discovered or revealed (notice how revealing the truth implies it was covered over, hidden, or obscured previously). The result of scientific discovery is supposedly positive knowledge—facts that are true. Charles Taylor calls this the "subtraction story"—once the metaphysical baggage and superstitious nonsense is "subtracted" away, we are left with the facts.[4] This kind of account positions science as a kind of liberator of truth and positions prescientific understandings or forms of knowledge as either wrong, partially true or outright misleading.

Science helps us progress beyond old "superstitions" and "myths" toward the truth.

Scientists and philosophers themselves often narrate what they do in this fashion. Here is Galileo denigrating the humanists that had gone before him:

> If what we are discussing were a point of law or of the humanities, in which neither true for false exists, one might trust in subtlety of mind and readiness of tongue and in the greater experience of the writers, and expect him who excelled in those things to make his reasoning most plausible, and one might judge it to be the best. But in the natural science, whose conclusions are true and necessary and have nothing to do with human will, one must take care not to place oneself in the defense of error.[5]

Notice how Galileo is positioning science as the appropriate method to access truth. The humanist tradition he presents as being driven by rhetoric and lacking certainty. The focus on total certainty was also a concern of Descartes and Hobbes. Both thought the intellectual traditions of their times were inadequate and sought radical new beginnings, seeking "a foundation for reconstruction of society and knowledge alike."[6]

Living during the highly destructive period of the Thirty Years War left Descartes and Hobbes anxious for something more definitive than the rhetorical and scholastic debates that also left Galileo disappointed. Descartes reports that

> from my childhood, I have been familiar with letters; and as I was given to believe that by their help a clear and certain knowledge of all that is useful in life might be acquired, I was ardently desirous of instruction. But as soon as I had finished the entire course of study, at the close of which it is customary to be admitted into the order of the learned, I completely changed my opinion. For I found myself involved in so many doubts and errors, that I was convinced I had advanced no further in all my attempts at learning, than the discovery at every turn of my own ignorance.[7]

Like Galileo, Descartes was disillusioned with the past and the state of thinking at his time. He was seeking a new beginning, and he was clear about this. His famous *Meditations* begins with him sitting in front of the fire and committing to completely start over in order to find out if there is anything that can be found that cannot be doubted and can therefore serve as a definitively sure foundation for knowledge. If such a foundation could be found and sure knowledge obtained, this knowledge would be powerful and compelling. Mary Midgley points out that while this is ostensibly an epistemological question it is also a "claim to rule."[8] Knowledge is power and has political consequences. This connection was clear to Descartes and Hobbes and thus finding a new foundation for knowledge was pivotal to remaking society to be more rational.

I am trying to draw attention to the attempt to break with the past and how Descartes and Hobbes were creating a narrative that vilified the past in order to justify their the sociopolitical and epistemological goals. This denigration of the past leads to an approach that Midgley calls "slum clearing"—the suggestion that for something new to be built the old has to be smashed down.[9] She claims that this language of completely getting rid of the old to make possible a new and secure foundation has been a much reused metaphor in modernity (it was in wide use in twentieth-century philosophy of language).[10]

Observe a telling comment from Descartes in the *Discourse on Method*:

> These ancient cities that were once mere straggling villages and have become in the course of time great cities are quite commonly poorly laid out compared to those well-ordered towns that an engineer lays out on a vacant plane as it suits his fancy. And although, upon considering one-by-one the buildings in the former class of towns, one finds as much art or more than one finds in the latter class of towns, still, upon seeing how the buildings are arranged—here a large one, there a small one—and how they make the streets crooked and uneven, one will say that it is chance more than the will of some men using their reason that has made them thus.[11]

Descartes's comment here is oddly aesthetic: "the old city has a certain charm, but wouldn't it be better and more beautiful if it had been designed rationally?" The lingering hint here is that we could, if we wanted to, perhaps knock it down and create a rational city.

The point I want to make here is that early modern figures such as Descartes and Hobbes created narratives about premodernity and modernity that radically separated the two and denigrated the former. Science is treated as a kind of savior that can liberate us from the chaos and confusion of old traditions.[12]

This sort of narrative shapes how the death of the cosmos is understood. In fact, the common "death of the cosmos to universe" narrative seems to fit Taylor's subtraction story perfectly: once Plato's ancient metaphysics and old religious traditions were rejected and rational scientific methods applied to understanding the cosmos, science discovered the truth which is the universe as we now know it. New technologies resulted in the discovery of the true nature of the universe. Like it or not, this is the truth that we are left with when we get rid of those old superstitions (as I mentioned in the introduction, when asked about whether it is a loss to lose the cosmos as a home, many of my students adopt a kind of Stoic pride in the Truth revealed by science, no matter what existential costs it has).

But this narrative is misleading on a variety of levels. On one level, it presents the break between premodern and modern scientific understandings too radically and too simplistically. Historians and philosophers of science have noted that it was not as if there was no evidence against the immutability of the cosmos previous to Copernicus and Brahe. New stars had appeared. Comets had been seen seemingly passing through the spheres. These and other irregularities did not appear to astronomers as things that meant the Platonic model was wrong, instead they could be written off as outliers, mistakes, or, as in Ptolemy and other astronomers, incorporated into the system by adding small spheres on larger spheres, and so on, in an attempt to account for strange movements. Once Plato's model was established, the general trend was to explain irregularities *into* the model, not to challenge the model itself.[13] In other words, to use the language of Thomas Kuhn, once the Platonic model became the paradigmatic for western thought, irregularities had to fit the model, not challenge it.[14] Only when there were too many irregularities, too many contradictions, could an outsider challenge the paradigm.

Simply put, the common narrative about the death of the cosmos is overly simplistic and, among other things, doesn't take into account the continuities between premodernity and modernity. For my purposes here, I want to resist the facile dismissal of the Platonic cosmos in order to do justice to how radical and important this shift was for our understanding of human nature, ethics and *amor mundi*. It is too easy, superficial and morally inattentive to just throw it away with the rest of old metaphysical baggage.

With this antagonism toward old, supposedly obsolete ideas in mind (and the critiques of such antagonisms and the narratives that ensued), let me move forward in discussing the changes that occurred with the death of the cosmos. I will begin with the metaphysics, then move to the epistemological, ethical, and anthropological shifts that followed.

FROM COSMOS TO UNIVERSE

So how is the (modern) universe different from the (Platonic) cosmos? The most obvious difference is the title from Alexandre Koyre's classic work on the topic *From the Closed World to the Infinite Universe*. Koyre calls the premodern cosmos a "closed world" in the sense that the extent of the cosmos is knowable and limited. From the Earth, at the center of the cosmos, we are able to see the whole cosmos around us. The universe, on the other hand, is infinite and expanding. It is incomprehensibly large and we are in it . . . somewhere. The scale of the cosmos is so different from the scale of the universe that the very language we use about it had to be changed:

> The long use of *world* to mean an object so patterned and unified as the geocentric *kosmos* went so far as to disqualify it as a name for what we now call "universe." When once the old model is broken, we become aware that a detailed knowledge of the Region of all regions will be forever unattainable. We now need a world not for some specific and imaginable object but foe "whatever the totality may be," and *universe* is such a word. Furthermore, the same process which broke the old model threw a new meaning on *world* [that of "inhabited realm"] which, while it lasted, made it impossible without grave inconveniences, to call the totality the *world.*[15]

The universe became so much larger, so unknown and so distant that it could no longer be called world. To call it "universe"—which basically means "all the stuff"—is to stress its vastness and alien-ness. What is out there? The term also reflects that the universe lacks the order that gave the cosmos its name. Whatever patterns we may find in the universe lack a coherent organization. The whole isn't an order—certainly not a teleological or anthropocentric one. There is no connection to human nature or ethics—it just is.

The metaphysical distinction in Plato between being and becoming is also discarded in favor of the distinction between material and nonmaterial. Here the demands of scientific epistemology seem to have bled over into ontology. When understanding the universe is a matter of scientific examination, this requires dealing only with the physical aspects of the universe that can be quantified and subject to verifiable experimentation. Anything spiritual, teleological, or meaningful (in the sense of "mattering" for humans) is disallowed in order to achieve objectivity. While these things are placed aside epistemologically, this has actually changed how many in modernity understand ontological reality: nonphysical aspects are not "real" in the sense that they are "subjective" or "cultural." You can find meaning, spiritual or otherwise, in the universe if you want, but it is personal or cultural belief and not capable of being measured, manipulated, and recreated by science (despite the desire to do so by some social scientists).

Positivism and early twentieth-century philosophy in general have been very dismissive of anything that is not "objective." In so doing, they are following "the Cartesian promise of absolute epistemic objectivity and ultimate foundations for knowledge."[16] Anything that is a matter of spirituality or human meaning is a matter of speculation, tradition, or opinion. Facts require the elimination of any such thing that would compromise objectivity. Metaphysically, this leaves a very bare bones material universe with no room for human meaning, spirituality, or purpose. Such is the price, we are told, of truth.

Thus, the positivist interpretation of Wittgenstein's famous final line of the *Tractatus*—"Whereof one cannot speak one must remain silent"—is that things that we cannot verify, such as ethics, meaning, and spirituality, are not philosophically relevant because we cannot say anything objective about them. In the Anglo-American philosophical tradition ethics fell into disrepute. After G. E. Moore, the questions in ethics became meta-ethical concerns about the status of ethics: "Can ethics be done at all?" and "What would that look like?"—until seventy years later when John Rawls shocked everyone by being crazy enough to just do ethics.[17]

I discussed this shift from a living, porous, enchanted cosmos to the modern disenchanted universe briefly in the last chapter. Keeping in mind Taylor's critique of the "subtraction story," the common articulation of this shift is that once the metaphysical baggage was stripped or "subtracted" away, we are left with a mechanical universe. Surrounded by a machine-like objects, we find ourselves in a universe calling out for study, tinkering, and fixing.[18] If some don't like this remaining mechanical universe, this is unfortunate because it's the reality we found once we subtracted away the rest.

As I have mentioned, this modern knowledge that the universe is an infinite, nonteleological, physical expanse is reflected in the language my undergraduate students use to describe it. When I ask them the first words that come to mind when I say "cosmos," they use words like "vast," "empty," "cold," and "infinite" that describe the universe. When I ask them why they don't say "home," quite a few express a kind of Stoic pride in the "truth" of a universe that doesn't care about humans and in which we randomly inhabit the "third rock from the sun."

Critics have worried that this empty, mechanical universe has lost the wonder and meaningfulness of the cosmos.[19] It is interesting to note how some scientists have attempted to recover a sense that science reveals the "wonders" of the universe. For example, a recent documentary series that was intended to raise science literacy in America was entitled *Cosmos* and not, say, *Our Cold, Meaningless Universe.*

FROM LOVE TO MODERN SCIENCE

So where is the love? If the universe is not orderly or rational in the sense that the cosmos was, then Plato would say it makes no sense to imitate or love the universe. And in fact, to moderns, it makes no sense to imitate the universe—it just is. Loving the cosmos, which involved observing it, basking in its glory, studying, imitating, and connecting with it, is no longer the way to know it.

Like early Plato, whom Nussbaum presented as rejecting emotion and love, modern science has little or nothing to say about love. It is not relevant to scientific epistemology. Perhaps some scientists might stress the joys and love of scientific practice, but one can do good science perfectly well in a curmudgeonly mood.

Richard Rorty famously argued that modern philosophy rejected metaphysics in favor of epistemology.[20] Or, stated differently, modern thinkers wanted to get epistemology sorted out before it said anything about the nature of reality. Only once we have the secure method Descartes sought, can we begin to discover what reality is like. But this is misleading; since modern scientific epistemology only accepts certain forms of "data" this means that only that part of human experience which can fit through this filter can be accepted as real.

In the *Discourse on Method*, Descartes outlines an atomistic vision of finding the smallest pieces from which everything else is composed. Just as one can take a car apart to see how it fits together, so too we could methodically take the world apart and figure out what it is made of. Once we know the basic parts and how they relate, then we could potentially fix parts that break, improve things, rearrange them, and so on. Arendt claims that one of the under-recognized aspects of science is the ability to recreate the phenomena in question. To truly understand is to be able to create. This turns reality into a kind of Lego universe which has basic building blocks that, once understood, can be potentially rearranged as we desire. And such knowledge is connected to power:

> Francis Bacon, in the sixteenth century, was the first to foresee the physical power potential in scientific knowledge. We are the first, I am suggesting, to have enough of that power actually at hand to create new possibilities almost at will. By massive physical changes deliberately induced, we can literally pry new alternative out of nature. The ancient tyranny of matter has been broken, and we know it. We found, in the seventeenth century, that the physical world was not at all like what Aristotle had thought and Aquinas had taught. We are today coming to the further realization that the physical world need not be as it is. We can change it and shape it to suit our purposes.
>
> Technology, in short, has come of age, not merely as technical capability, but as a social phenomenon. We have the power to create new possibilities, and the will to do so. By creating new possibilities, we give ourselves more choices. With more choices, we have more opportunities. With more opportunities, we can have more freedom, and with more freedom we can be more human. That I think, is what is new about our age. We are recognizing that our technical prowess literally bursts with the promise of new freedom, enhanced human dignity, and unfettered aspiration. Belatedly, we are also realizing the new opportunities that technological development offers to make new and potentially big mistakes.[21]

I quote this disturbing example of modern technological optimism at some length to point to some of the key assumptions at work here. The connection between science and technology is clear and exciting: thanks to science, technology, engineering, math (STEM), we are more free, have more choices, and more dignity than ever before. Some of the language here is antagonistic: Mesthene says we can now be "free of the tyranny of physical nature."[22] We saw in the last chapter that Plato and other ancient philosophers denigrated the physical nature, including the human body, emotions, love, sexuality, and hunger in favor of the intellectual. Here the limitations of physicality are treated as tyrannical and being able to control physical nature and alter it indefinitely to fit our smallest whims is presented as deeply liberating and dignifying. The Earth, the cosmos, and our bodies are not mentioned specifically, but as woefully physical things, we can only assume they too will be sloughed off or altered thanks to our new phenomenal cosmic powers. Only tacked on at the end is some recognition that sometimes this can go wrong and we should probably be careful.

Overall, the vision here is an odd mix of resentment, maniacal excitement for power, and commitment to human dignity. It is tempting to say that this is an extreme example of technological optimism, but I am not sure that it is unusually extreme. Modern capitalist societies tend to be deeply Baconian in this regard and quick to seek technological solutions to any and every problem. We can see here how modernity is a project with radical goals not only to understand reality but also to engineer it. Knowledge provided by science is the key to this project.

Early thinkers of modernity such as Bacon, Descartes, and Hobbes agreed that math was the key to doing good science. In so doing they followed Plato yet again. Jonas claims there is a clear line of descent from Plato's demiurge in the Timaeus to Leibnitz's mathematically inclined God.[23] When the cosmos becomes the universe, it becomes quantifiable and quantified: "Kepler was probably the first of the moderns to declare quantity (or magnitude) to be the, at once, essential and truly knowable aspect of reality: cognition accordingly consists to him in measurement and comparison of measurements."[24] Interestingly, Jonas points out that the kind of math being used by modern thinkers was different from that of the ancient Greeks. Moderns were using "algebra applied to geometry" instead of the "classical geometry" of the Greeks because the moderns' "paramount interest in motion, as against the satisfaction with pattern, which prompted the ascendance of algebraic method in physics: motion instead of fixed spatial proportions became the main object of measurement. This marks a radically novel attitude. In the early stages of modern science, analysis of becoming supplants contemplation of being."[25] For the ancients, math was about appreciating pattern. For the moderns, math was about dynamics and movement. Ancient math has a more aesthetic component related to the emphasis on contemplation. Modern math seeks to understand in order to practically interact and intervene in nature.

These differences are stark enough that Jonas concludes: "Indeed, a different 'nature' could be interrogated by a different mathematics."[26] Based on what we are seeing, this is correct. These are different maths, aimed at different natures or realities—one the inspiring cosmos, the other the quantifiable, alterable universe.

MODERN ETHICS

With the death of the cosmos, ethics was undermined. No longer could one look up to the stars and find ethics there—ethics had lost its place. For the early modern thinker, one could still attempt to anchor ethics to God, but, as we will see, the death of the cosmos also disrupted traditional belief in God.

Not surprisingly, just as moderns sought for a new foundation to knowledge and a new beginning to society, they also sought for new ethics, especially a scientific ethics. Observe how Sidgwick describes Utilitarianism as a rational systematization of human activity:

> If we are not to systematise human activities by taking Universal Happiness as their common end, on what other principles are we to systematise them? It should be observed that these principles must not only enable us to compare among themselves the values of the different non-hedonistic ends which we have been considering, but must also provide a common standard for comparing these values with that of Happiness.[27]

Putting aside the technical difficulties of the hedonistic calculus, the promise of Utilitarianism, according to Sidgwick, is to create a "scientifically complete and systematically reflective form of that regulation of conduct."[28] Sidgwick explicitly seeks a scientific ethics that can regulate conduct and he thinks this can only be done with a quantitative method that approximates science. And while no ethical system has been able to achieve the status of scientific fact, it has not been for lack of trying. Even rival thinkers to Utilitarianism like Kant sought for a new ethics that would have a secure foundation.

Rémi Brague argues that technology becomes the morality of modernity:

> Modern cosmology neutralizes the world. The difference between good and evil nevertheless remains. Ancient and medieval nature, inasmuch as it was offered up for contemplation in celestial phenomena, was the reign of the good; modern nature, present above all in biological facts, appears to be the kingdom of evil. But the good comes back to nature: since good is not in nature, it is thus necessary to introduce it into nature. And by force, by taking nature against the grain . . . Modern technology defines itself through an undertaking of domination, through a plan to become, according to the famous epigram of Descartes, the

> 'master and possessor of nature.' Technological activity was considered up until the modern era as a perfection of nature. It was a matter of delivering nature of that which it could not produce by itself. . . . Henceforth, it was a question of imposing external order upon nature. If technology could set out to ameliorate nature, it was because nature left a lot to be desired.[29]

Brague says that the ancients and medievals recognized the value of technology, but did not imbue it with the moral significance that moderns tend to: "For the Moderns, on the other hand, to battle nature was to battle evil and spread good."[30] What is interesting about Brague's claim is how the universe has been neutralized and emptied of inherent value—it is neither good nor bad, it just is—and yet it is reinscribed into a moral story that treats it as evil. The ontological neutrality actually leads to it being treated as a moral evil in need of correction.

The Earth and nature are placed into an antagonistic relationship with humans: improving the human condition means controlling and manipulating nature. With regard to love, benevolence toward humans seems to require a decrease in love of the Earth and nature. Environmentalist thinkers have long pointed to the anthropocentrism in modern thought, but what is surprising is the resentment and occasional hatred that is targeted at nature. In the Mesthene comments, technology is located as the savior that releases us from the tyranny of nature. The only nature that is bearable is apparently a nature that is completely fungible and made available for whatever human desires occur to us. The love that Plato expressed for the cosmos is completely replaced here with anger and resentment.

And this anger at nature can also be applied to the Earth itself. Arendt begins *The Human Condition* with a discussion of the exuberant reactions to the launch of the first satellite in 1957 that claimed this marked the first "step towards escape of man's imprisonment on earth."[31] She points out that "future man, whom the scientists tell us they will produce in no more than a hundred years, seems to be possessed by a rebellion against human existence as it has been given, a free gift from nowhere (secularly speaking), which he wishes to exchange, as it were, for something he has made himself."[32] Instead of gratitude and love for the Earth and nature, which are the conditions of our continued existence, modernity seems to resent that there are any conditions to human existence at all.

MODERN HUMAN NATURE

This frustration with the "tyranny" of nature and the Earth can also be aimed at traditional cultural practices, religion and "irrational" humans. The project of modernity seeks to remake society to make it more rational and this involves remaking humans themselves.

According to Castro-Gómez and Foucault, the goal in this regard is to produce a new rational kind of human: *homo economicus*.[33] This kind of human is supposed to be calculatively rational, working through their preferences and making rational decisions to bring about their desired goals. In order to create a positive image of this new modern man, premodern or nonmodern humanity needed to be characterized as backward, primitive, dirty, and irrational. In terms of specific practices to bring this about, Castro-Gómez claims this was done in Latin America through the use of constitutions, etiquette manuals, and grammar manuals. All three of these created an image of modern "civilized" man: a constitution by legally defining who counts as a citizen and what benefits they get, etiquette and grammar manuals by delineating what behavior and what kind of language is civilized. Civilized man is held up by casting others down: he is rational, modern, and civilized; they are irrational, primitive, and barbaric.

This use of "civilized" versus "savage" is not new. James C. Scott points out that this kind of narrative was used by agricultural societies in an effort to tempt huntergatherers into working on farms.[34] But the modern effort to make the new rational man took place against the backdrop of the death of the cosmos. As an ongoing project, modernity has been aggressive about reshaping humanity. The Enlightenment was explicitly named to emphasize the freedom it offered from not just the tyranny of the Earth and nature, but also religion, tradition, and primitive cultures. To become modern and enlightened is to leave all this behind.

Historically, this process has involved both a racial and gendered component. As Genevieve Lloyd has pointed out, the man of reason was explicitly the *man* of reason. She shows how western thought has long associated men with rationality and women with emotion.[35] When Descartes seeks for "clear and distinct" ideas, this involves being free from emotions. Since women are associated with emotions, this can bring doubt as to whether women can be rational at all and potentially serve as a justification for excluding them from many activities that require rationality. But it also means that to be a rational man means to be less "womanly." There is an ethical ideal of a disconnected and distanced autonomous individual who can judge objectively, and this idea is tied to masculinity.

This same sort of characterization has happened with race also, as a certain race was associated with rationality and therefore modern civilization, while others were associated with irrationality and savageness. African Americans in the United States were stereotyped as more sexual and more "animal" than "rational" whites.[36] In Latin America, strict racial categories and nomenclature arose to mark how European or how "Indian" a person was. To this day to call someone an *indio* is considered very insulting in many parts of Latin America. These racist and sexist characterizations remain in modern society, and it has been a long, protracted, and ongoing effort to understand how they influence modern institutions, modern cultural attitudes, and contemporary social problems.[37]

If erotic love and emotions are exiled from rationality, and cast down onto women, people of color, and premodern communities, then "civilized" modern men are only left with the kind of cold, rational *amor mundi* that Plato seemed to extoll. This puts limits on the power of *amor mundi* for modern man, but there is another aspect of modern identity that has contributed even more to the decline of *amor mundi* and this is modern atomistic individualism.

The modern concept of human nature tends to radically separate the individual from the world and from other individuals. In the last chapter, I discussed the difference between what Charles Taylor calls the premodern "porous" self and the modern "buffered" self in order to clarify how the premodern person was more open and connected to the world around them than the modern individual. I also discussed how the premodern world was, to use Weber's famous phrase, an "enchanted" cosmos.[38] It was full of spiritual and physical beings that were directly connected to and could influence every person. The actions of spirits, other human beings, and animals could affect me and my actions could affect them on a level that was much more profound than what we moderns are accustomed to. My choices about something so simple as what to eat, when, with whom, and where, could anger or appease the gods, the spirits, my ancestors, animals, and so on. To say that the modern individual is buffered is to say that our modern understanding of what is it so be human radically separates us from this traditional porous connection to the cosmos, nature, and other humans.

A related point that helps illustrate the changes in how human nature is understood in modernity concerns the "disembedding" of the self. By the phrase "embedding," Taylor is drawing attention to the way that a person can be so immersed and suffused with a particular place or condition that they cannot be separated from it. He claims that there have been at least two major phases of disembedding, the first is what Karl Jaspers called the "Axial Age"—the period in which many of the world major religious and intellectual traditions, such as Daoism, Confucianism, Zoroastrianism, and Greek philosophy arose. The second is with the project of modernity. Previous to these disembeddings, humans were embedded at least three ways: socially, within the cosmos, and in "existing reality."[39] To be embedded socially is to be so thoroughly tied to a particular community that you cannot relate to the world or God except through and with that community. In extreme cases, it would mean having no separate identity from the group. To be embedded in the cosmos is to be located in a specific place that is suffused with meaning and includes sacred places. The landscape itself would be experienced as full of history, meaning, and spirituality. Last, to be embedded in an existing reality refers to how pre-Axial people tended to not separate out a spiritual realm from this physical one. They experienced this reality as being both spiritual and physical at the same time.

With the Axial Age, many religions or intellectual traditions began to pose the existence of a separate realm from this one, whether the realm of spirituality, the gods or something like Plato's forms. If one is seeking for the intellectual or spiritual, this may involve a certain renunciation of this realm in favor of the desired alternate realm. This may also involve a break with the other humans in this realm or escaping some part of this cosmos for another part (or another place entirely).

Of course, it is in modernity that disembedding reaches a fever pitch. Interestingly, as modernity became more secular, there was an increasing focus on "existing reality" as accessed by science, but that existing reality was increasingly purged of anything spiritual or meaningful. The physical universe was all that was left; meaning, ethics and significance were no longer treated as things that were out in the world, capable of being found. Rather morality and meaning were instead "in the mind" of human beings.[40] Furthermore, the "buffered" modern human was no longer as intimately tied to society or the cosmos as they were in premodernity.

Taylor calls this kind of modern hyperindividualism "political atomism."[41] He means by this that modern political theory tends to start with atomistic individual human beings and ask how society can be created and function from all these individuals. Taylor's choice of the term "atomism" seems a good choice to describe this modern approach to understanding human society, because it points us to the Cartesian desire to break things into their smallest component parts in order understand how they relate together. This works the other way too however, as the atomistic modern individual is posited as the basic building block and then the question is how do these individual parts make society.

If the modern universe is approached as a kind of Lego universe that can be deconstructed and reconstructed, then so too human society can be deconstructed and reconstructed. Modern political theory, according to Taylor, often begins with the assumption that this basic building block is the human individual.[42] Taylor points out the fascination that early modern political thinkers had with stories like Robinson Crusoe, the lone man trying to survive by himself on an island. He also points out that some modern theories of language also tend to start with two individuals meeting and trying to figure out how to communicate. Hobbes' social contract theory treats the members of the contract "as if but even now sprung out of the earth, and suddenly, like mushrooms."[43] Of course, Descartes famously argued that after doubting everything, the one thing that was left was my own mind (who knows if I actually have a body?).

Taylor argues this modern obsession with the atomistic understanding of human nature is mistaken. There is no atomistic individual. When Descartes attempts to doubt everything and claims to end up with only the (individual) human mind, he fails to doubt language, the medium that is necessary to communicate his claims, and language is not individual. We learn it in a community with others. Another way to point to the same problem, as Eva Kittay points out, is to remember that we all have mothers. We are not mushrooms that come from nowhere, we come from other people.[44]

I will not delve into the full critique of modern hyperindividualism that thinkers like Taylor and Kittay are making, but I do want to stress a further point, which is the ethical appeal of this vision of human nature as atomistic. Taylor argues that this kind of atomism has been thoroughly debunked and critiqued. For good philosophical reasons it should no longer be attractive—and yet it is. Why?

Taylor claims that despite the shaky philosophy behind it, this atomistic model remains powerful precisely because it is not just an ontological theory about the nature of humans, but an ethical image of humans. In other words, even though it has little ontological reality, it remains popular because people want it to be true—they like the ethical ideal that is working here. And what is this ethical ideal? According to Taylor, it is the ideal of rational self-control and autonomy.[45] First of all, the buffered self is safer than the porous self. The spirits, gods, and other humans that might wreck my plans are either disenchanted away or separated out into other autonomous human beings (and, if we follow Descartes, who knows if they exist?). Safely distanced from all that, I am now in a position to control my own life and take responsibility for my own choices. Thus, there is also freedom in being buffered and disembedded. There is a certain masculine dignity to being autonomous, in control, and responsible. Additionally, it can be tied to the epistemic goals of objectivity.

Arendt argues that there is a less noble motivation in this early modern ideal of autonomy: escape from social responsibility.[46] While post-Axial religions and intellectual traditions may have posited alternative realms and sanctioned some degree of rejection of human society in order to seek spiritual or intellectual goals, premodern society, nonetheless, did not have the radical individualism of modernity. Individuals in a society were expected to take responsibility for the entire society to some degree. If you didn't perform the appropriate rituals at the appropriate time, you might put all of us at risk, and as such we have a responsibility to make sure you fulfill your responsibilities. This kind of mutual checking could obviously become onerous. These sorts of social rules also put limits on economic relations. Christians, for example, were not supposed to make money through interest (one reason why Jews, who were allowed to charge interest as long as it was with non-Jewish people, were often the bankers).[47] Furthermore, the Old Testament criteria for the success or failure of a religious community is the condition of the widow or orphan.[48] To be wealthy and living well, while the widow and orphan are dying, is a sin since we are all brothers and sisters.

Early capitalists had to overcome these prohibitions in order to engage in the kinds of economic practices that have become normal now and this meant, in part, breaking the explicit connection between the individual capitalist and the widow and the orphan. In other words, it meant finding ways to erase the moral responsibility toward the widow and the orphan.[49]

Just as creating the image of a rational civilized man requires putting down emotional women and uncivilized savages, creating the space for modern individual responsibility and autonomy seems to require cutting away traditional social responsibilities. This is not to claim it has to be this way and many modern thinkers would claim traditional societies failed in many ways that modernity has improved. But in terms of *amor mundi*, this cutting loose of the modern self from the cosmos, world, and society is not an innocent move but a highly complex one that has deep implications. We may object against Professor Higgens, who sings in My Fair Lady that

Who prefers to spend the evening in the silence of his room
Who likes an atmosphere as restful as an undiscovered tomb[50]

The real world is too human and too noisy! This quiet might be nice for him of course, but who is washing his clothes while he thinks and writes in the study? Who watches the children? The well-written irony of this part of his misogynistic song is that his complaints about "humanity's mad inhuman noise" actually reflect the inhumanity of his own buffered distance from humanity.[51]

I have attempted to discuss some of the ontological, epistemological, ethical, and anthropological shifts that occurred with the death of the cosmos. Just as ontology, epistemology, ethics, and humans were connected in the Timaean cosmos, so too they are interconnected and related in the modern universe. The atomistic understanding of the universe is also applied to human society and human relations. The ethical ideals of autonomy are tied to the epistemological ideals of objectivity. Obviously, these are immensely complicated shifts and they have been discussed and debated by many important thinkers. For our purposes here, my goal has been to attempt to show some of the relevant aspects of this shift that help explain the decline of *amor mundi* in modernity.

The disembedding and buffering of the modern self disconnects us from the cosmos, the Earth and other humans. The ideals of autonomy and self-responsibility turn us inwards. The disenchantment of the cosmos has resulted in a mechanical universe which we are called to understand and "fix" where necessary. This relationship is not about love, but control and mastery. At times, it even seems to be based on resentment and hatred. The necessity of objectivity to discover the laws of the universe requires the rejection and control of emotions, including love. So we find ourselves in a universe which has no morality or inherent meaning (these are in the human mind) and which has no teleological purpose. We can control and fix the universe through objective, rational control. Is it any surprise that the project of modernity has seen the decline of *amor mundi*?

The death of the cosmos has radical implications for ontology, epistemology, ethics, and human nature. Before moving on, I want to examine three thinkers who dwell on the *death* part of the death of the cosmos. Each of these thinkers (Hans Jonas, Carolyn Merchant, and Friedrich Nietzsche) can help us see important features of life after the death of the cosmos.

HANS JONAS: THE ONTOLOGICAL DOMINION OF DEATH

Jonas is an important thinker for understanding the death of the cosmos. The role it plays in his thought shows that he understood the profound power of life in predmodernity and sought to awaken a sense for it in his contemporaries. In *The Phenomenon of Life*, he claims that life was omnipresent in the premodern world. Looking around us at the world we see animal and plant life everywhere and even inanimate objects were often thought of as living beings. Jonas points out that when life is everywhere around us, the exception and oddity is death. Life was not a question. It was reality. "Death, not life, calls for an explanation."[52] Death, as the strange unreality, has to be explained in terms of life.

Modern thought, on the other hand:

> is placed in exactly the opposite theoretic situation. Death is the natural thing, life the problem. From the physical science there spread over the conception of all existence an ontology whose model entity is pure matter, stripped of all features of life. . . . The tremendously enlarged universe of modern cosmology is conceived as a field of inanimate masses and forces which operate according to the laws of inertia and of quantitative distribution in space. This denuded substratum of all reality could only be arrived at through a progressive expurgation of vital features from the physical record and through strict abstention from projecting into its image our felt aliveness.[53]

Instead of the panpsychism of premodernity, now we find a modern "panmechanism"—all is machine. And if we interpret everything around us in mechanical terms, then life becomes the phenomenon that is difficult to explain: "To take life as a problem is here to acknowledge its strangeness in the mechanical world which is *the* world."[54]

This brings Jonas to a remarkable conclusion:

> The earlier goal, we have seen, was to interpret the apparently lifeless in the image of life and to extend life into apparent death. Then, it was the corpse, this primal exhibition of "dead" matter, which was the limit of all understanding and therefore the first thing not to be accepted at its face-value. Today the living, feeling, striving organism has taken over this role and is being unmasked as a *ludibrium materiae*, a subtle hoax of matter. Only when a corpse is the body plainly intelligible: then it returns from its puzzling and unorthodox behavior of aliveness to the unambiguous, "familiar" state of a body within the world of bodies, whose general laws provide the canon of all comprehensibility. To approximate the laws of the organic body to this canon, i.e., to efface in *this* sense the boundaries between life and death, is the direction of modern thought on life as a physical fact. Our thinking today is under the ontological dominance of death.[55]

This is a powerful description of the difference between the premodern living cosmos and the dead modern universe. The language of death is clearly important for Jonas's purposes and his claim that modern thought is under the "ontological dominance of death" is wonderfully provocative. A mechanical universe carries connotations that some might find worrying, but a universe of corpses is terrifying.

Jonas thinks the problem of life and the living body ought to be at the center of philosophical concern—both because of the ethical implications that follow from a dead universe, but also at the level of ontology and epistemology. He thinks modern thought adopted this mechanical model out of hopes that it would enable a better understanding of the world, but that ultimately the corpse universe obscures as much as it illuminates: "waiving the intelligibility of life—the price which modern knowledge was willing to pay for its title to the greater part of reality—renders the world unintelligible as well. And the reduction of teleological to mechanical causality, great as its advantages are for analytic description, has gained nothing in the matter of comprehending the nexus itself: the one is no less mysterious than the other."[56] In other words, the mechanical model does give us power over objects, but it must empty reality of meaning, purpose, and life to do so. And since we are living beings who care about meaning and purpose, it closes off a fundamental aspect of life in order to understand it. As such, he concludes, it fails to understand.

This complaint that modern ontology and epistemology are reductive and make our studies of human experience inadequate is one that is shared by

Jonas' teacher Heidegger, his teacher Husserl, and other German thinkers of their time. But science has proven so successful in giving us technological control, that it has become hegemonic in academia. The humanities and their concern with human meaning are increasingly treated, as we saw with early-twentieth-century Anglo-American ethics, as unverifiable and merely a matter of preference.

Jonas sees existentialism as a movement trying to cope with the death of the cosmos and finding itself in a corpse universe: "Gone is the cosmos with whose immanent logos my own can feel kinship, gone the new order of the whole in which man has a place. That place appears now as a sheer and brute accident."[57] Meaning and value are no longer something we can find in the world, but something we must choose: "A universe without an intrinsic hierarchy of being, as the Copernican universe is, leaves values ontologically unsupported, and the self is thrown back entirely upon itself in its quest for meaning and value."[58] Throughout his works, Jonas complains of "this estrangement between man and the world," but he does not think that existentialism's emphasis on individual choice can overcome this estrangement.[59] Like Arendt and later Heidegger, overcoming world alienation for Jonas requires finding a way to give the world back its dignity. For Jonas, this means reaffirming and philosophically rethinking and respecting the life in the cosmos.

Jonas seeks to reapproach the phenomenon of life through a philosophical biology. He offers an "'existential' interpretation of biological facts."[60] This is an attempt to extend the existential concerns that affect humans to all of life. Simply put, if the rest of life is truly alive, then it too cares about its existence. For Jonas, humans have a responsibility to help protect life and foster it, but the mechanization and death of the cosmos has alienated humans from the world in a way that has greatly impoverished our ability to recognize this task—much less carry it out.

Jonas is, in his way, calling us to love the world again. I will discuss later how he can help us think about this task. I want to turn now to a second thinker on the death of the cosmos, Carolyn Merchant.

CAROLYN MERCHANT: THE DEATH OF NATURE

Carolyn Merchant argues that the scientific revolution led to the death of nature. On the surface, this might not seem to be significantly different from the claims of Jonas. But Merchant emphasizes a gendered component. The death of nature is the death of *Mother* Nature.

What is the significance of the gender of nature? We have already seen earlier in this chapter that the project of modernity and key early modern thinkers followed the tendency in ancient Greek philosophy to treat emotion, love and the physical body as dangers to rationality and knowledge. Becoming rational and civilized meant overcoming physical desires and emotions. Irrational emotion, sexuality, and brut physicality were assigned to women and certain (newly constructed) racial groups. In this sense, to become rational was a masculine, racialized task, and it was doubtful whether women or Africans or Native Americans could do so.

Merchant argues that nature itself was commonly personified as a woman. The language of "Mother Nature" is still used today at times. To conceive of nature as our mother requires certain kinds of relationship with her and the metaphors here end up being very important. It may sound odd to modern ears, but the early modern poets John Milton, Edmund Spencer, and John Donne all compared mining to rape. To dig in the Earth and extract desired minerals or gold was an act of violation. We can take from the Earth what she generously gives, but we should not tear her open to get what we want:

Men also, and by his suggestion taught,
Ransack'd the Center, and with impious hands
Rifl'd the bowels of thir mother Earth
For Tresures better hid. Soon has his crew
Op'nd into the Hill a spacious wound
And dig'd out ribs of Gold.[61]

Thus, the metaphor of mother Earth served to prohibit certain forms of behavior and relations with the Earth. Furthermore, Merchant argues that the world was understood to be an organism. In other words, it was alive. As such, it could respond to us depending on how we interacted with it. As we just discussed with Jonas, the world was alive, and that it was alive required it be related to as a complex, living thing. Those who angered it—or her risked suffering her wrath.

At the same time that some thinkers were invoking gendered understandings of Mother Earth as a reason to defend her, other thinkers were using gender as a justification for various forms of attack. Machiavelli, for example, characterized fortune as a violent woman:

> Fortune is the ruler of half our actions. . . . I would compare her to an impetuous river that when turbulent, inundates the plains, casts down tress and buildings, removes earth from this side and places it on the other; everyone flees before it and everything yields to its fury without being able to oppose it; and yet though it is of such a kind, still when it is quiet, men can make provision against it by dikes and banks, so that when it rises it will either go down a canal or its rush will not be so wild and dangerous. So it is with fortune which shows her power

where no measures have been taken to resist her, and directs her fury where she knows that no dikes or barriers have been made to hold her.[62]

He stresses that she must be subdued:

For fortune is a woman and it is necessary if you wish to master her, to conquer her by force; and it can be seen that she lets herself be overcome by the bold rather than by those who proceed coldly, and therefore like a woman, she is always a friend to the young because they are less cautious, fiercer, and master her with greater audacity.[63]

Merchant claims that alongside gendered metaphors that treat nature as a loving, organic mother there arose gendered metaphors that stressed disorderliness, chaos, and wild sexuality. The embodiment of this was the witch, a dangerous threat who had to be defeated through cunning and the application of strength/power.[64]

Early modern scientists invoked sexual language to describe finding out nature's secrets: "but likewise for the further disclosing of the secrets of nature. Neither ought a man to make scruple of entering and penetrating into these holes and corners, when the inquisition of truth is his whole object."[65] The womb, previously treated as the metaphorical location of minerals and gold, is now treated as the location of secrets that the scientist must find. In one case, a prohibition on rape, in the second, a seeming justification for rape. Alternatively, finding the secrets of nature could replace the language of breaking into the womb with the language of declothing nature and exposing her nakedness.[66] Here the language does not involved rape, but seems to sexualize nature and encourage seducing her to see her naked "truths."

For those who weren't willing to rape in the name of science, the metaphor had to be changed. Merchant argues that mechanical metaphors began to replace the traditional organic metaphors. If nature is a woman, then mining and seeking secrets from her could be treated as akin to rape. But if nature was a complex machine, then there was no ethical prohibition about intervening in it. In fact, if *it*—not *her*—is a machine, we probably *should* intervene, since machines often need fixing or improvement.

Nature, whether woman or machine, needs domination. But when the organic order of the premodern cosmos is replaced by the dead modern mechanical order, intervention is less ethically fraught. Either way, the fundamental attitude toward the Earth and nature changes from respect, care, gratitude, and love to control, domination, and power. Mother Earth is comprehensively loveable. Earth as machine could never have the same ethical status as a mother.

While Jonas emphasizes the death of the cosmos and seeks to restore dignity to it and even love for it, Merchant shows some of the complex gender issues that are involved. Her argument shows how the Earth has been gendered as a woman and this has shaped human relations with her: Mother Earth demands respect and caution—at least until some early modern scientists explicitly sought to find her secrets. Merchant argues that the ways that gender and sexual metaphors are invoked by predominantly male early modern scientists is not without significance and she powerfully shows how these seem to tie in with a (sometimes sexualized) desire for domination. As the metaphors switch from organic living ones to dead mechanical ones, this domination is further justified.

Like Jonas, Merchant thinks the death of nature is clearly connected with the rise of modern science. I want to turn now to one further thinker who agrees and extends the scope of modern death.

NIETZSCHE: THE DEATH OF GOD

In aphorism 125 of *The Gay Science*, Nietzsche famously declares that God is dead. This is one of the more egregiously misunderstood passages of philosophy. Here is how it begins:

> Have you not heard of that madman who lit a lantern in the bright morning hours, ran to the market place, and cried incessantly: "I seek God! I seek God!"—As many of those who did not believe in God were standing around just then, he provoked much laughter. Has he got lost? Asked one. Did he lose his way like a child? Asked another. . .—Thus they yelled and laughed.[67]

The madman announces the death of God. But notice that it is not believers who respond to him, but the unbelievers—"those who did not believe in God." And their reaction is to mock him. To them, it is silly to be concerned about the death of God. God was just a story, who cares that he is gone? The madman then says that we have killed God:

> All of us are his murderers. But how did we do this? How could we drink up the sea? Who gave us the sponge to wipe away the entire horizon? What were we doing when we unchanged this earth from the sun? Whither is it moving now? Wither are we moving? Away from all suns? Are we not plunging continually? Backward, sideward, forward, in all directions? Is there still any up or down? Are we not straying as through an infinite nothing? Do we not feel the breath of empty space? Has it not become colder? Is not night continually closing in on us?[68]

The madman is telling these unbelievers that they do not understand the radical implications of the death of God—otherwise, they would not find this funny. It is clear from the passage that the death of God does not refer only to the decline of belief in the Christian God. Nietzsche describes being lost without direction in the new empty and cold universe. He describes the earth being unchained from the sun. He is describing the death of the cosmos and all the corresponding losses—including the creator of this orderly cosmos. Instead of being located and sheltered in a cosmos created by God, we find ourselves without a God in a meaningless universe. The death of the cosmos, in turn, implies the death of traditional order and the narratives that helped to make sense of this life (what Nietzsche calls "myth"), the death of traditional ethics, and the end of the intellectual sway of Plato. Almost two thousand years of tradition crumbling away.

Nietzsche was very critical of Christianity. But by having the madman speak to "those who did not believe in God," he is trying to wake us all up to the fact that the death of the cosmos changes everything. The madman, after giving his speech and seeing astonishment and lack of comprehension from his listeners, throws down his lamp and declares he has come too early. This is Nietzsche saying that we do not understand the implications of the death of the cosmos and the death of God. Everything has changed and we do not even realize it yet.

The Timaean Cosmos was a meaningful cosmos. The modern universe is not. Speaking of science, Nietzsche says:

> What? Do we really want to permit existence to be degraded for us like this—reduced to a mere exercise for a calculator and an indoor diversion for mathematicians? Above all, one should not wish to divest existence of its *rich ambiguity* . . . an interpretation that permits counting, calculation, weighing, seeing, and touching and nothing more—that is a crudity and a naiveté, assuming that it is not a mental illness, an idiocy. . . . A "scientific" interpretation of the world, as you understand it, might therefore still be one of the *most stupid* of all possible interpretations of the world, meaning that it is one of the poorest in meaning...an essentially mechanical world would be an essentially *meaningless* world. Assuming that one estimated the *value* of a piece of music according to how much it could be counted, calculated and expressed in formulas: how absurd would such a "scientific" estimation of music be! What would one have comprehended, understood, grasped of it? Nothing, really nothing of what is "music" in it![69]

For Nietzsche, modern science empties the cosmos of meaning. For most humans, this is a disaster: regular people are left without the "instincts" that allowed them to make sense of the rich ambiguity of life. In his early work, Nietzsche was intensely focused on the myths which, in *The Birth of Tragedy*, he calls "the concentrated picture of the world . . . abbreviature of phenomena."[70] The world is a very complicated place and there are many ways of making it intelligible. Myth simplifies the world and establishes a tradition for a particular people. Nietzsche says that "without myth, however, every culture loses its healthy creative natural power: it is only a horizon encompassed by myths that rounds off to unity a social movement."[71]

Different traditions or communities have different myths and different instincts. Nietzsche characterizes Christianity as one such tradition that established a certain myth or set of myths as expounded in the Bible, in the stained glass windows of Cathedrals, in stories of particular saints, in communal practices, and so on. These myths expounded a certain view of the world, of the self, of how one should act, of the nature of time, and so on—all these became embodied as instincts, the taken-for-granted ways of being and understandings that guide a community. Despite his many reservations about Christianity, Nietzsche recognizes that it made evil meaningful, gave life a certain purpose, and as such fended off nihilism for a period of time.[72] Every human community or tradition has a guiding set of myths that help to organize a set of instincts—thus we speak of the Confucian tradition, the Buddhist tradition, the Islamic tradition, and so on—each with many variations.

The death of God is the death of traditional myths, which were ways to make sense of the world. For Nietzsche, this is a problem because the death of God(s) leaves us without direction:

> Let us now think of the abstract man unguided by myth, the abstract education, the abstract morality, the abstract justice, the abstract state: let us picture to ourselves the lawless roving of the artistic imagination, unchecked by native myth: let us imagine a culture which has no fixed and sacred primitive seat, but is doomed to exhaust all its possibilities, and to nourish itself wretchedly on all other cultures—there we have the Present, the result of Socratism, which is bent on the destruction of myth. And now the mythless man remains eternally hungering amid the past, and digs and grubs for roots. . . . Let us ask ourselves whether the feverish and uncanny excitement of this culture is anything but the eager seizing and snatching at food of hungry man.[73]

And:

> The tremendous historical need of our unsatisfied modern culture, the assembling around one of countless other cultures, the consuming desire for knowledge—what does all this point to, if not to the loss of myth, the loss of the mythical home, the mythical maternal womb?[74]

Modernity often narrates itself as a project seeking to replace old myths with a new rational society, a new ethics, and new ways to organize socially. But Nietzsche thinks these efforts have failed. Instead of creating a coherent modern myth, Nietzsche claims in some places that we have a cacophony of often incoherent and incompatible myths.[75] In other places he seems to argue we have new bad myths that fail to allow for human flourishing. Of course, both could be true. While patriotism and nationalism are common "failing myth" targets for Nietzsche, he also constantly rails against capitalism and the "money earning creature":[76]

> This expansion belongs to the most beloved of the dogmas of modern political economy. As much knowledge and education as possible; therefore the greatest possible supply and demand—hence as much happiness as possible:—that is the formula. In this case utility is made the object and goal of education,—utility in the sense of gain—the greatest possible pecuniary gain.[77]

A money grubbing, utility maximizing society is, for Nietzsche, a pathetic one. It remains empty at heart:

> What I combat is economic optimism: as if increasing expenditure of everybody must necessarily involve the increasing welfare of everybody. The opposite seems to me to be the case: expenditure of everybody amounts to a collective loss: man is diminished—so one no longer knows what aim this tremendous process has served. An aim? a new aim?—that is what humanity needs.[78]

His name for the meaningless emptiness of the new universe and the vacuum this leaves for humans is nihilism. Like Jonas, Nietzsche thinks modern science gives us a dead universe. The dead universe is emptied of meaning and modern attempts to fill in these holes are paltry.

For Nietzsche, the universe of modern science is bad ontology and bad epistemology. It is modern science imposing on reality their own myth: the myth that the universe is mechanical and fully comprehensible.[79] To make such a claim fails to appreciate the richness of the cosmos/universe. Without delving into his own theory of the metaphysics of reality,[80] he rejects the idea that the universe is mechanical and *only* mechanical. In his later works, he repeatedly excoriates modern thought for being cold, insect-like (inhuman), and insensitive to the deep importance of human meaning:

> All great problems demand *great love*, and of that only strong, round, secure spirits who have a firm grip on themselves are capable. It makes the most telling difference whether a thinker has a personal relationship to his problems and finds in them his destiny, his distress and his greatest happiness, or an "impersonal" one, meaning that he can do no better than to touch them and grasp them with the antennae of cold, curious thought.[81]

Since neither science nor capitalism offer a meaningful life, Nietzsche pointed out that people would be susceptible to all kinds of narratives that offered such a promise; desperate for meaning, we end up unfulfilled and "snatching at food." So how can nihilism be overcome?

Interestingly, Nietzsche's answer is love.

AMOR FATI: LOVE OF FATE

In his early works (especially under the influence of Wagner), Nietzsche seeks to overcome nihilism through the founding of a new myth. In *The Birth of Tragedy*, he is particularly impressed by Greek myth and Greek tragedy. Greek tragedy provided a communal solution: it created immersive stories that tied people together as a community while preserving the full gamut of human experience (it is important for Nietzsche that we not pretend we can eliminate the ugly and painful parts of life, but instead find ways to make sense of them).[82] I think the community-focused approach of his early philosophy is a kind of *amor mundi*, and we could add it to our list of ways to love the world. For our purposes here however, I am interested in his discussion of *amor fati* in his later works.

In *The Gay Science*, the same work that announces the death of God, Nietzsche says the following:

> I want to learn more and more to see as beautiful what is necessary in things as what is beautiful in them—thus I will be one of those who make things beautiful. *Amor fati*: let that be my love from now on! I do not want to wage war against ugliness. I do not want to accuse; I do not even want to accuse the accusers. Let looking away be my only negation. And, all in all on the whole: someday I wish to be only a Yes-sayer![83]

This is how he announces the task of *amor fati*—loving fate. What does he mean by this? To love fate is to love all aspects of life, including the ugly and painful ones. This can be done in many ways: (1) by creating another place that is the "true reality"—heaven for Christians or the realm of intellectual perfection for Plato—that is supposedly perfect and free from ugliness and pain, (2) by claiming ugliness and pain are not real and attachment to them or the illusion of them can be dispelled through ascetic practice (Stoicism or Buddhism), (3) by eliminating evil and suffering insofar as is possible and improving the human condition (the project of modernity), and so forth. Nietzsche thinks that (1) is a fiction that is no longer believable, (2) is life denying and (3) cannot succeed because no matter how brilliant a future we may create, there is always a painful past that made it possible (no Hegelian future will justify the "slaughterbench of history").

The ugly and painful part of life cannot be eliminated. They must be accepted. To love fate is to love life, with all the good and ugliness that this entails:

> But what if pleasure and displeasure are so intertwined that whoever *wants* as much as possible of one *must* also have as much as possible of the other—that whoever wants to learn to 'jubilate up to the heavens' must also be prepared for 'grief unto death'?
>
> Even today you still have the choice: either as little displeasure as possible, in short, lack of pain . . . or as much displeasure as possible as the price for the growth of a bounty of refined pleasures and joys that hitherto have seldom been tasted. Should you decide on the former, i.e., if you want to decrease and diminish people's susceptibility to pain, you also have to decrease and diminish their capacity for joy.[84]

To have the highest highs, we must embrace it all, including the lowest lows. This means that we must learn to "not want to wage war against what is ugly." Instead, we must learn to love: "Love, too, has to be learned."[85]

Nietzsche proposes a kind of mental experiment that will help us figure out if we love life, the so-called "myth" of eternal recurrence:

> What if some day or night a demon were to steal into your loneliest loneliness and say to you: 'This life as you now live it and have lived it you will have to live once again and innumerable times again; and there will be nothing new in it, but every pain and every joy and every thought and sigh and everything unspeakably small or great in your life must return to you, all in the same succession and sequence—even this spider and this moonlight between the trees, and even this moment and I myself. The eternal hourglass of existence is turned over again and again, and you with it, speck of dust!' Would you not throw yourself down and gnash your teeth and curse the demon who spoke thus? Or have you once experienced a tremendous moment when you would have answered him: "You are a god, and never have I heard anything more divine." If this thought gained power over you, as you are it would transform and possibly crush you; the question in each and every thing, "Do you want this again and innumerable times again?" would lie on your actions as the heaviest weight! Or how well disposed would you have to become to yourself and to life to long for nothing more fervently than for this ultimate eternal confirmation and seal?[86]

To love life is to embrace it, all the joy and wonder along with all the ugliness and pain, even if it were repeated over and over. For Nietzsche, this seems to be a matter of will, a choice that each of us can make. We can embrace the task of learning to love life and it can make life richer—both more painful, but also more joyous. Thus, the "gay" part of *The Gay Science* which we would now probably translate as "joyous wisdom" or "joyous knowledge."

On the surface of things this seems to be an individualistic task. Each of us has the responsibility to seek to love our own fates. Certainly, most contemporary readers of Nietzsche interpret him this way.[87] And it is on this register that we can ask if *amor fati* solves the problem of nihilism. *Amor fati* as Nietzsche presents it rejects the ontological dominion of the mechanical universe but doesn't provide a new myth. It recognizes that reality is ambiguous and rich without needing to simplify it into a new "mythical womb." After describing the death of God as an event that is immensely dangerous to most humans who need meaning, Nietzsche claims that for the select few "free spirits," it represents a new opportunity: the opportunity to explore and seek knowledge without the narrow constraints of a particular set of meanings. This is the "science" or knowledge part of the "gay science."

The free spirit who learns to love the many different ways of understanding the world and move through different meanings likely does overcome nihilism for themselves. If anything, the free spirit now has access to an excess of meaning. This could be dangerously overwhelming, but it would not be nihilism. The free spirit understands the scientific-mechanical universe without reducing all of reality to it. Thus, she could also understand and experience the world as a living cosmos.

If *amor fati* is successful at solving nihilism it seems it solves the individual problem of nihilism, but not the social problem of nihilism that motivated Nietzche's early work. And Nietzsche is clear in his later works that learning to love in this way is a task that is only for the few. Most humans will likely still need myth. Julian Young argues that the social problem of nihilism requires the establishing of a new myth and that the free spirits in the later works are potential founders of such a new myth.[88] On this reading then the later concern with free spirits is also a potential solution to the social problem of nihilism. This is something of an idiosyncratic reading of Nietzsche but one I think has much merit. But the standard reading of Nietzsche sees the later focus on free spirits as an individualistic focus.

Both Heidegger and Arendt seem to read Nietzsche this way and were critical of him for this individualism and his focus on individual will to power. While Heidegger criticizes Nietzsche for his focus on will, Arendt articulates *amor mundi* as a foil to the individualism of Nietzschean *amor fati*. Even though Nietzsche is talking about loving the world, his approach makes this an individualized task of loving the world—it is my fate in this world that I must learn to love. Arendt shifts the focus from *fate* to *world* in an effort to show that learning to love is not just the task of an individual. Furthermore, she has a specific meaning of world that makes her claim different from Nietzsche's quest to love life.

In many ways Arendt's analysis parallels Nietzsche's. Like Nietzsche and Heidegger, she will present our dangerous current philosophical and social situation as a product of a long decline. Like both of them, she claims there is something at the beginning of the Western tradition that has been lost and her work seeks to remind us of it. But Arendt is not convinced that *amor fati* can overcome nihilism—at least not the *social* problem of nihilsm, so she proposed her own account of nihilism and her own alternative, *amor mundi*. In the next two chapters, I will seek to explore her account of the loss of *amor mundi* and what it would mean for us to learn to love in this way again.

This moment ends the primarily historical focus of this book. Up to this point I have attempted to show, in chapter 1, how *amor mundi* (love of the cosmos) was considered one of the highest forms of human activity in Western thought. In this chapter, I have explored how the death of the cosmos and the rise of modern science led to a decline and forgetting of *amor mundi*.

From this point in this book, my focus will be on forms of *amor mundi* that can and ought to be recovered. I do not think *amor mundi* in the Platonic sense of loving the cosmos can be recovered because the Platonic cosmos is irrecoverable after Copernicus. This does not mean that there is not some form of love of the cosmos that could be a live option for humans today, but rather that it would have to show that how the modern universe is meaningful instead of an empty, vast expanse or propose an alternative cosmos.[89] I am not going to delve into this option and instead I will claim that Arendt presents an important form of *amor mundi*—love of a world—that is recoverable and that Native American philosophy also offers an important form of *amor mundi*—love of Earth/nature/place—that is also recoverable. I will spend the next two chapters on Arendt and one on the thought of Vine Deloria Jr. and Daniel Wildcat trying to make that case. Perhaps by developing a love for a world and the Earth we could also learn to love the cosmos.

NOTES

1. Gillespie, Michael Allen. *The Theological Origins of Modernity* (Chicago: University of Chicago Press, 2008).

2. Taylor, Charles. *A Secular Age* (Cambridge, MA: Harvard University Press, 2007).

3. Brague, Rémi. *The Wisdom of the World: The Experience of the Universe in Western Thought* (Chicago: University of Chicago Press, 2003), 186–187.

4. Taylor, *A Secular Age*.

5. Quoted in Ravetz, Jerome. "From Descartes to Rumsfeld: The Rise and Decline of Ignorance of Ignorance" in Matthias Gross and Linsey McGoey, eds.

Routledge International Handbook of Ignorance Studies (New York: Routledge, 2015), 54.

6. Ravetz, "From Descartes to Rumsfeld," 54.
7. Ibid., 53.
8. Midgely, Mary. *Wisdom, Information and Wonder: What Is Knowledge for?* (New York, Routledge, 1989), 23.
9. Midgely, *Wisdom, Information and Wonder*, 27.
10. Ibid., 27.
11. Descartes, Rene. "Discourse on Method," in Roger Ariew and Eric Watkins, eds. *Modern Philosophy* (Indianapolis, IN: Hackett Publishing, 2009), 34.
12. Midgley has also point this out in Midgley, Mary. *Science as Salvation: A Modern Myth and its Meaning* (New York, Routledge, 1994).
13. https://www.princeton.edu/~hos/mike/texts/ptolemy/ptolemy.html. Accessed August 17, 2018
14. Kuhn, Thomas. *The Structure of Scientific Revolutions* (Chicago: University of Chicago Press, 1996).
15. Quoted in Brague, *The Wisdom of the World*, 189.
16. Bordo, Susan. *The Flight to Objectivity: Essays On Cartesianism and Culture* (Albany: SUNY Press, 1987).
17. For a critique of Moore and twentieth-century Anglo-American philosophy in this regard, see MacIntyre, Alasdair. *After Virtue* (Notre Dame, IN: Notre Dame Press, 2007).
18. Merchant, Carolyn. *The Death of Nature: Women Ecology and the Scientific Revolution* (New York: Harper and Row, 1990). See later in this same chapter.
19. This is one of critiques made by the Romantics. I will not go into detail on this point here however.
20. Rorty, Richard. *Philosophy and the Mirror of Nature* (Princeton: Princeton University Press, 1981).
21. Mesthene, Emmanuel G. "The Social Impact of Technological Change" in Robert C. Scharff and Cal Dusek, eds. *Philosophy of Technology* (Malden, MA: Wiley Blackwell, 2014), 681–682.
22. Mesthene, "The Social Impact of Technological Change," 680.
23. Jonas, Hans. *The Phenomenon of Life: Toward a Philosophical Biology* (Evanston, IL: Northwestern University Press, 2001), 66.
24. Jonas, *The Phenomenon of Life*.
25. Ibid., 67.
26. Ibid., 68.
27. Quoted in Nussbaum, Martha. *The Fragility of Goodness: Luck and Ethics in Greek Tragedy and Philosophy* (New York, NY: Cambridge University Press, 1989), 112.
28. Nussbaum, *The Fragility of Goodness*, 113.
29. Brague, *The Wisdom of the World*, 210.
30. Ibid.
31. Arendt, Hannah. *The Human Condition* (Chicago: University of Chicago Press, 1998), 2.

32. Arendt, *The Human Condition*, 2–3.

33. See Castro-Gómez, Santiago. "The Social Sciences, Epistemic Violence and the Problem of the 'Invention of the Other.'" *Views from South*, Vol. 3, No. 2, 2002, pp. 269–285 and Foucault, Michel. *Discipline and Punish* (New York: NY: Vintage Books, 1995) and Foucault, Michel. *Power/Knowledge* (New York: Vintage, 1980).

34. Scott, James C. *Against the Grain: A Deep History of the Earliest States* (New Haven: Yale University Press, 2017).

35. Lloyd, Genevieve. "The Man of Reason." *Metaphilosophy*, Vol. 10, No. 1 (January 1979).

36. Brandt, Allan M. "Racism and Research: The Case of the Tuskegee Syphillis Study." *Hastings Center Report*, Report 8, no. 6 (Dec. 1978): 21–29.

37. Mills, Charles W. *The Racial Contract* (Ithaca: Cornell University Press, 1997) and Sullivan, Shannon and Nancy Tuana, eds. *Race and Epistemologies of Ignorance* (Albany: SUNY University Press, 2007).

38. Weber, Max. *Economy and Society* (Berkeley, CA: University of California Press, 1978).

39. Taylor, Charles. *Dilemmas and Connections* (Cambridge: The Belknap Press, 2011), 217–221.

40. Taylor, *Dilemmas and Connections*, 219.

41. Taylor, Charles. *Philosophical Papers: Volume 2, Philosophy and the Human Sciences* (New York, NY: Cambridge University Press, 1985).

42. Taylor, *Philosophical Papers*.

43. Quoted in Kittay, Eva Feder. *Love's Labor: Essays on Women, Equality and Dependency* (New York: Routledge, 1999), 23.

44. Kittay, *Love's Labor*.

45. Taylor, *A Secular Age*.

46. Arendt, *The Human Condition*.

47. Hyde, Lewis. *The Gift: Imagination and the Erotic Life of Property* (New York: Vintage, 1983).

48. Heshel, Abraham. *The Prophets* (New York: Perennial Classics, 2001).

49. Arendt, *The Human Condition*.

50. My Fair Lady, 2001.

51. Smith, Dorothy. *The Everyday World as Problematic: A Feminist Sociology* (Boston: Northeastern University Press, 1989).

52. Smith, *The Everyday World as Problematic*, 8.

53. Ibid., 9–10.

54. Ibid., 11.

55. Ibid., 12.

56. Ibid., 25.

57. Ibid., 214.

58. Ibid., 215.

59. Ibid., 216.

60. Ibid., 1.

61. Merchant, *The Death of Nature*, 39.

62. Ibid., 130.

63. Ibid., 130.
64. Ibid., 132.
65. Ibid., 168.
66. Ibid., 190–191.
67. Nietzsche, Friedrich, *The Gay Science* (New York, NY: Vintage, 1974), aphorism 125.
68. Nietzsche, *The Gay Science*, 1974.
69. Ibid., aphorism 373.
70. Nietzsche, Friedrich, *The Birth of Tragedy* (New York, NY: Dover, 1995), 85.
71. Nietzsche, *The Birth of Tragedy.*
72. Nietzsche, Friedrich. *The Will to Power* (New York, NY: Vintage, 1968), 9–10; see also Nietzsche, Friedrich, *The Genealogy of Morals* (New York, NY: Oxford University Press, 1999), 108–114.
73. Nietzsche, *The Birth of Tragedy*, 85.
74. Ibid., 85.
75. There is a scholarly debate about whether Nietzsche wants to embrace the fragmentation of modernity or laments it. At least in his early writing, like *The Birth of Tragedy*, he seems to lament it.
76. Nietzsche, Friedrich. "On the Future of Our Educational Institutions," in *The Complete Works of Friedrich Nietzsche, Vol 3* (Edinburgh, Eng: Morrison and Gibb Limited, 1910), 37.
77. Nietzsche, "On the Future of Our Educational Institutions," 136.
78. Nietzsche, *The Will to Power*, 866.
79. *The Birth of Tragedy* calls this second claim "Socratism."
80. See Nietzsche, *The Will to Power.*
81. Nietzsche, *The Gay Science*, 1974, Aphorism 345
82. Julian Young offers a compelling account of early "communitarian" Nietzsche. See Young, Julian. *Nietzsche's Philosophy of Religion* (New York: Cambridge University Press, 2006).
83. Nietzsche, Friedrich. *The Gay Science* (New York, NY: Cambridge University Press, 2001), 276.
84. Nietzsche, *The Gay Science*, 2001, 12.
85. Nietzsche, *The Gay Science*, 1974, 334.
86. Nietzsche, *The Gay Science*, 2001, 341.
87. Julian Young does not and I find his claims compelling. See Young, *Nietzsche's Philosophy of Religion.*
88. Ibid.
89. There are scientists that are exploring this area. See for example Barad, Karen. *Meeting the Universe Halfway: Quantum Physics and the Entanglement of Matter and Meaning* (Virginia: Duke University Press, 2007).

Chapter 3

Arendt and *Amor Mundi*

What Is World?

In the first chapter, I explored how a particular kind of *amor mundi*—love of the cosmos—was considered of the highest joys and intellectual activities by some in premodern Western thought. Typically, it is contemplation and not love that is given these accolades, but if we locate contemplation in the context of the Timaean cosmos, we see that contemplation has to be aimed at the proper target: the orderly, magnificent cosmos (and, in Judeo-Christian-Islamic traditions, God the creator of the cosmos). Understood properly, contemplation is not just thinking but seeking a deep connection with the cosmos—in short, a kind of cosmic love.

If this is correct, then the question becomes what happened to cosmic *amor mundi*? I attempted to answer this question in chapter 2 and claimed that the rise of modern science undermined the ancient model of the cosmos, resulting in radical changes to Western ontology, epistemology, ethics, and anthropology. The replacement of the meaningful, living, ethics-bestowing cosmos with the mechanical universe left contemplation without a target and humans without a home. We are still dealing with the fallout today.

In this chapter and the next, I want to turn to Hannah Arendt. In so doing, I am turning away from the historical account of the rise and decline of cosmic *amor mundi* toward other conceptions of world and the question of how we can find meaning after the death of the cosmos. This was a pivotal concern of early-twentieth-century German thought.[1] For Arendt, this question came to her through Nietzsche and her teachers Heidegger and Jaspers. In the reading of Arendt that I will present, she concurs with these thinkers that the hegemony of positivist science has choked off the ability to address concerns of human meaning and purpose—science and the logical positivists of the time sought to only deal with that which could be quantified, measured, and verified—and on everything else one was supposed to remain silent. But this meant that with regard to the matters that were most salient to human existence (What is the meaning of life? Why are we here? What is wrong and right? What should I do?), science had nothing to say. Insofar as other disciplines sought to emulate science and its success, they also had nothing to say. It was in this context that existentialism rebelled and sought to put these questions first and foremost.

However, Arendt, like Heidegger, worried that existentialists put too much emphasis on the individual. Like later Nietzsche, the existentialists tended to treat nihilism as an individual problem. The social problem of nihilism often remains in the background. As we saw in the last chapter, Nietzsche in his later works seeks to overcome nihilism through *amor fati*—love of (one's) fate. For Arendt, this answer is not adequate, and she proposes an alternative, *amor mundi*, which addresses nihilism as a social problem.

There are some problems that make this task complex however. The first problem is that Arendt never wrote a systematic treatment of *amor mundi*. She suggested *Amor Mundi* as the title for the book that later was called *The Human Condition*. Her editor thought the phrase *amor mundi* was not clear enough, however, and suggested the current title. This presents something of a mystery: what is the significance of Arendt considering the title *Amor Mundi* for what we now are familiar with as *The Human Condition*? If the text had been entitled *Amor Mundi*, the reader might expect a phenomenological account of a particular kind of love—love of the world. This would be a unique and compelling topic. But this is not what we get in *The Human Condition*. Indeed, love is not a prominent topic in it. Love is mentioned in five different places in the text, and the most extensive discussion, which is still rather short, concerns love as an anti-political force. Why, then, would Arendt think that *Amor Mundi* would be an appropriate title?

Despite these difficulties, the claim that Arendt is concerned with *amor mundi* has been an attractive one to Arendt scholars. It serves to locate and define her as an interesting foil to Nietzsche's *amor fati* and a worldly rebuttal to later Heidegger's more solitary and meditative approach to philosophy.[2] Read in this light, she emerges with a different account of nihilism and a kind of social optimism about potential ways to respond to it. It is not surprising that Elizabeth Young-Bruel gives her biography of Arendt the subtitle "For love of the world."[3] Bernauer even argues that *amor mundi* serves as a kind of faith for Arendt, as a visionary articulation of a potentially better world.[4]

I agree with these scholars that *amor mundi* is a powerful, underappreciated, and undertheorized part of Arendt's philosophy. My claim is that if you do accept the importance of *amor mundi* for Arendt, then some aspects of *The Human Condition* begin to look strange. If anything, the text focuses on the opposite of *amor mundi*: world alienation. This completely changes the tone of the book—it is not a book that feels alive with love. At most, perhaps, there is a quiet longing for this love that is lost. Like many of Arendt's works, it is in some ways a somber book, tracking with a critical eye the genealogy of Western world alienation, exposing in the end a consumerist and atomistic society that does not even recognize what it has lost. As such, also like many of Arendt's works, it is tragic—in this case a double tragedy: first the loss, then the forgetting. What Arendt is doing is, in a sense, a remembering. As a good phenomenologist she starts here and now in our condition of world alienation and by alerting us to this tragedy, by thinking it through, seeks to remind us of a possibility we have lost and forgotten. Perhaps, then, *amor mundi* is omnipresent in *The Human Condition*, but like a ghost—present in its absence.

This means that to present *amor mundi* as part of a solution to the death of the cosmos will require reconstructing what she means by it in positive terms. This is the work of these chapters. The focus of this chapter is what Arendt means by "world." The focus of the next chapter will be what she means by "love" of a world.

The second problem is that by *amor mundi* she means something different from the kind of love of the cosmos that has been discussed in chapters 1 and 2. Furthermore, Arendt also offers an alternative account of modern nihilism, one that is in key ways quite different from the account I have offered so far but that also has parallels with it. This means that there are multiple forms of *amor mundi* which vary depending on what is meant by "world." The differences are important enough that the phenomenology of each form will be quite distinct, but they are similar enough that they can be grouped together as a foil to modern, more individualistic conceptions of love.

Let me make a note here about how I am approaching the topic. I am presenting the topic of *amor mundi* in this book in a different order than my own intellectual exposure to it. I came to think about the topic through Arendt first, then realized Native American thinkers I was reading were also critiquing Western thought for its alienation and lack of love toward the Earth. I finally realized that *amor mundi* made a powerful lens to understand the shift from the living cosmos to the dead universe that was being discussed by Brague, Jonas, Nietzsche, and others. I have chosen to begin with love of the cosmos to show that *amor mundi* is not an obscure issue of Arendtian scholarship, but fundamental to the development of Western thought and contemporary social and moral problems. Furthermore, I present love of the cosmos first because I think of the three forms of *amor mundi* I will be discussing, it is the most difficult to recover. The death of the cosmos and decline of love of the cosmos helps us to see aspects of our experience that are important to understand our times, but the cosmos cannot simply be resurrected—at least not on the Platonic model. Love of a world (Arendt) and love of the Earth/Nature/place (Deloria and Wildcat) on the other hand are "live" options—we can begin immediately to learn to love the world on these terms. Perhaps in doing so, we can expand this love to the cosmos.

EARTH AND WORLD

So what does Arendt mean by world? To help clarify this, let me begin by looking at the distinction Arendt makes between Earth and World.

In the prologue to *The Human Condition*, Arendt begins by pointing out some of the disturbing language invoked after the launch of the Soviet satellite Sputnik in 1957. She claims that "the immediate reaction, expressed on the spur of the moment, was relief about the first 'step toward escape from men's imprisonment on Earth.'"[5] She claims this comment, made by an American reporter, was not a slip of the tongue, but a common sentiment that could be summarized by the following statement which was carved on the gravestone of a Russian scientist: "Mankind will not remain bound to the Earth forever."[6]

Seeing this sentiment she asks: "Should the emancipation and secularization of the modern age, which began with a turning-away, not necessarily from God, but from a god who was the Father of men in heaven, end with an even more fateful repudiation of an Earth who was the Mother of all living creatures under the sky?"[7] This fascinating passage is similar to Nietzsche's death of God aphorism, which stresses the superficial dismissal of God. Here Arendt worries about a superficial alienation from the Earth. Mother Earth is presented as a condition for the possibility of human life, a fundamental aspect of human existence—and yet she/it is resented. Indeed, Arendt claims this reflects "a rebellion against human existence as it has been given, a free gift from nowhere (secularly speaking), which he wishes to exchange, as it were, for something he has made himself."[8]

This opening salvo is striking and Arendt sees it as part of a "boomerang" effect of "science's great triumphs."[9] She goes on to talk about other effects of the rise of modern science, including how scientific language disrupts "political" language and how automation threatens a society obsessed with labor. She finishes these introductory comments with the claim that *The Human Condition* seeks to "trace back modern world alienation, its twofold flight from the Earth into the universe and from the world into the self, to its origins."[10] Seemingly, based on this passage, world alienation is related to Earth alienation. What is odd is that, despite its prominence in the prologue, Earth alienation plays little to no role in the rest of the text. "Earth" is not even included in the index. Thus, while Arendt begins the text with Earth alienation and includes it in her initial statement of the focus of the text, the discussion of the flight from the Earth to universe never materializes.

If *The Human Condition* is approached from a perspective that is primarily concerned about the Earth, this opening is very promising and exciting, but the rest of the text will likely be disappointing—world alienation, not Earth alienation is the focus of the text. Indeed, I will claim that Earth and Earth alienation are a primary concern for Wildcat and Deloria Jr., and that as such they might be frustrated with the failure of Arendt to delve more fully into these issues. It should be said, however, that this does not mean the text has nothing to teach us about Earth alienation. I will claim that the genealogy of world alienation which focuses on "the flight from the world into the self" has many parallels to what a genealogy of Earth alienation might look like—in other words the turn away from world to the self is also a turn away from Earth to the self. Arendt's genealogy of world alienation, in the end, will also illuminate Earth alienation.

WORLD

In order to clarify what is meant by "world," Arendt distinguishes between the private realm and the public realm. The private realm refers primarily to the home, a location that, in ancient Greece at least, was centered on activities of labor. Labor refers to activities like making food, cleaning, making clothing, and so on. These are activities that reflect our biological necessities and keep us alive and healthy. They are also very repetitive and often onerous tasks. In ancient Greece, according to Arendt, the private realm is kind of like the back stage where things are prepared. Having eaten, bathed, and dressed, those who have the luxury and freedom of being able to leave the private sphere and go on stage into the public sphere can do so. A fully human life, for the Greeks, is only possible if one can appear in public with others. And this public is the "location" of world:

> The term "public" signifies the world itself, in so far as it is common to all of us and distinguished from our privately owned place in it. This world, however, is not identical with the Earth or with nature. . . . It is related, rather, to the human artifact, the fabrication of human hands, as well as to affairs which go on among those who inhabit the man-made world together. To live together in the world means essentially that a world of things is between those who have it in common, as a table is located between those who sit around it; the world, like every in-between, relates and separates men at the same time.[11]

The kind of world Arendt is describing is a particular human world. Such a world refers, in part, to the buildings, roads, walls, and objects that physically make up a particular place. What these objects look like, how they are organized, how they relate to each other are all a part of that world. A village in South Korea will look physically different from a village in France. A world is recognizable to some degree by the architecture, organization, and style of that world.

But it also refers to the culture and the social meanings that develop between the humans that live in that place. Obviously, these cultural and social meanings affect how the physical objects in the world are built. In cities like Singapore where cultural beliefs in Feng Shui are prevalent, the physical buildings are built with quirks that reflect these beliefs—such as large holes in the middle of skyscrapers to allow for dragons to move through them or direct the flow of chi. As these cultural and social meanings are *built into* the place, they become embodied and will, in turn, exercise an influence back on new generations of humans that live there. Thus, there is a reciprocal relationship between the physical objects in the world and the sociocultural meanings and norms of the humans that interact in this world.

World, then, has an interesting ontological status. It does have a physical location, and it consists to some degree of the physical objects that make up a particular world, but it is also suffused with meaning by the history, narratives, and people who live in this world. If a particular world dies, its skeleton may be left behind in the form of ruins. We may learn about the "spirit" or "soul" of this world through its literature, philosophy, art, religions, and history. The language or languages that are spoken in a particular world are an intimate part of it and the norms and meanings of that world will be woven into that language in significant ways.

Importantly, any particular world transcends the humans that live in it. We are born into a world. It existed before us, and it will exist after us. For the individual raised in a world, it will appear far larger than themselves, reaching back into history to some foundation or past and possibly reaching into the future. The individual is a part of that world and immersed in it. And yet one could also leave one particular world and enter another. We still at times today talk about world in a way that reflects living in different, sometimes overlapping worlds: when a person is exposed to a community that is different from their own they might exclaim, "Wow, it is like a whole new world!" Museums often structure their exhibits according to different worlds. The Museo Nacional del Antropología in Mexico City is a rather stunning example of this, as a museum-goers pass in and out of different rooms that are each based on different indigenous groups. There is a large and very impressive room with artifacts and information about the Aztecs. Another room is about the Maya, another about the Toltecs, and so forth. There is a palpable sense that one is passing out of one world and into another as one passes through the rooms.

And yet this very real sense is also superficial and voyeuristic. It is simply not possible to truly experience a world in a museum. At best one can only get a very rough and inadequate image of these worlds. Film also serves as another form of voyeuristic world sampling—both in that film can show us different worlds (a documentary or a film set in a culture of country that one may not be familiar with) and even create new worlds (a fantasy or science fiction setting). The alternative worlds that are created in film and literature can never be fully fleshed out and rich as a real world and thus often function through a kind of illusion that one is looking into and experiencing a different world. And yet people often become highly invested in these worlds: the "world" of Harry Potter, the "world" of *The Lord of the Rings*, the "world" of *Star Wars*. That these worlds can take on a life of their own and surpass the original imagination of their creators is itself a fascinating phenomenon and the power of such worlds—even when they are clearly fantasies—does point out important features of "world."

This overlapping multiplicity of worlds is mostly a modern (or perhaps postmodern) phenomenon, reflecting the fragmentation of premodern worlds and the proliferation of modern worlds. In fact, I suspect Arendt would hesitate to give these fragmented modern worlds true status as worlds. She thinks that modernity, for reasons I will address shortly, is alienated from world. Indeed, as a result of our modern world alienation and our overlapping fragmented worlds, it may be hard for us to grasp the "depth" of a premodern world. This is a similar problem to the difficulties that modern thinkers might have understanding the "porousness" of the enchanted, living cosmos and how this can impede comprehending what is meant by "contemplation."

This once again presents the problem of how to understand (insofar as is possible) the premodern experience of world, in which a particular world would often be experienced as *the* world. Premodern societies were smaller, less mobile, and enchanted. For Arendt, the Greek city-state is a good example of a premodern world. For her, Athens is a particularly important example because it shows a kind of commitment to a world that embodies what is meant by politics or the political. Furthermore, many current Western political terms and traditions descend from ancient Greece: democracy, tyranny, monarchy, oligarchy, and so on. Thus, Arendt returns to Greece in an effort to understand the origins of Western political thought and also to recover aspects of it that have been forgotten. Let me turn to this story.

POLITICS AND WORLD

Arendt argues that there are important aspects of Greek understanding of politics and world that we have largely forgotten. A world, for the Greeks, was an accomplishment, almost a work of art. It was something that transcends humanity and shapes each new individual born into it. It connects us with each other and with past and future generations, all connected to a common world. A world is thus an inheritance—one that we are given and one that we can give to future generations: "Labor and work, as well as action, are also rooted in natality in so far as they have the task to provide and preserve the world for, to foresee and reckon with, the constant influx of newcomers who are born into the world as strangers."[12]

Note that Arendt's stress is not on *making* a world, but on *preserving* the world. A world, according to Arendt, is fragile. Living in a time of world alienation and overlapping fragmented worlds may make it seem like worlds can be invented and remade almost at will, but Arendt stresses the difficulty of preserving a world. This is a fairly conservative attitude, not in the political sense, but in the sense of *conserving* the world. This is not to say that Arendt thinks a world cannot or should not change—any particular world is always changing (thus older folks often complain that "everything has changed") and some particular world might have aspects that need to be changed (sexism or racism, etc.)—rather Arendt is claiming that there tend to be more forces attempting to pull a world apart than keep it together. We can see this in a striking comment she makes in an essay on education:

> The responsibility for the development of the child turns in a certain sense against the world: the child requires special protection and care so that nothing destructive may happen to him from the world. But the world, too, needs protection to keep it from being overrun and destroyed by the onslaught of the new that bursts upon it with each new generation.[13]

The first point she makes about protecting children from the world is one that every parent is aware of. We shield children from aspects of the world. Her second point is that the world itself needs protection from the "onslaught of the new." Each new child and each new generation will come into the world and even if they are taught to respect and care for the world, they will not take it over exactly as their parents did. Change is inevitable, and the world can handle some change. It is rapid, unconstrained, and overwhelming change that threatens to undermine and fragment a world.

So what does it take to "provide and preserve" a world? It takes labor, work, and action. Each of these terms is central to Arendt's analysis in *The Human Condition* and she spends a chapter on each. Briefly, as we have seen earlier, labor refers to those activities that are necessary for the biological existence of humans: growing crops, preparing food, making clothes, cleaning, and so on. Work refers to actually building the semipermanent objects that make up the physical "there-ness" of a world: roads, buildings, statues, public art, pottery making, carpentry, and so on. Work has the dignity of knowing that it is adding to and continuing the physical beauty of this world.

Action refers to the words and acts of those who live in the world. In terms of preserving the Greek world, it was specifically the actions of those who are free and allowed to be the citizens of this world that mattered most. These words and deeds help uphold this world, establish or continue its honor and dignity, and form its history.

Labor and work are fairly straightforward for our purposes here: labor keeps us alive, work builds the physical part of the world. The problem with labor and work for the Greeks (according to Arendt) is that they are not free. Freedom here means to be free from necessity: "To be free meant both not to be subject to the necessity of life or to the command of another *and* not to be in command oneself. It meant neither to rule nor to be ruled."[14] To work at a job from 8 am to 5 pm is not freedom. One cannot do what one wants, but rather must instead do what the bosses instruct. But the bosses themselves are not free—they must manage others instead of doing what they may want to be doing. For action to be free, it must be free from necessity, whether this is necessity to eat, or clean, or build roads. This means that not everyone can be free. The freedom of the few is predicated on the labor and work of the majority.

Only those who can leave the private sphere of labor and necessity can be citizens and join in the activities of the polis—the place where those who are free could gather. When we speak of Greek democracy, it doesn't mean everyone gets a vote. Only those who are free citizens can vote, and this was a minority of the population. In the United States, citizenship was originally only extended to those who were free from necessity (white landowning males). According to Greek understanding, these are the only people who are free and have the free time (*skhole*) that one needs to participate in the polis.

It is important to keep in mind that the goal of the citizens that met in the polis was not to govern the city-state of Athens—in other words, to manage it. Governing or managing a city is like being a boss, and it entails a lack of freedom. The whole point was to escape necessity and be free with others: "No activity that served only the purpose of making a living, of sustaining only the life process, was permitted to enter the political realm."[15] That they created a space for freedom was precisely what differentiated the Greeks, to their mind anyways, from other societies. What made other people barbarous was their lack of understanding of freedom.

We moderns tend to think of politics as managing society. Arendt thinks this completely fails to understand the meaning of politics and freedom. The *bios politikos*, the political life, was a kind of second life that was a possibility only for those that were citizens.[16] And it was in the polis that these free citizens had the opportunity to add to the legacy of the common world through speech and acts. A powerful speech, which when heard would be recorded or remembered, served as a form of world preservation. Such speeches make sense of what has happened in the past and is happening now in the world, creating a narrative that weaves together the past and the present and thus upholding and reinforcing this world. Great acts need to be retold and commemorated through speech to be made part of the "permanent record."

Being free of labor and work enable one to participate in politics, which in turn supports and glorifies the world. I should be careful not to make it sound like the polis was a calm, caring place—Arendt claims it was "permeated by a fiercely agonal spirit, where everybody had constantly to distinguish himself from all others, to show through unique deeds or achievements that he was the best of all."[17] The polis was a place of competition:

> The *polis* was supposed to multiply the occasions to win "immortal fame," that is to multiply the chances for everybody to distinguish himself, to show in deed and word who he was in his unique distinctness. One, if not the chief, reason for the incredible development of gift and genius in Athens, as well as for the hardly less surprising swift decline of the city-state, was precisely that from beginning to end its foremost aim was to make the extraordinary an ordinary occurrence of everyday life. The second function of the *polis* . . . was to offer a remedy for the futility of action and speech; for the chances that a deed deserving fame would not be forgotten, that it actually would become "immortal," were not very good.[18]

Thus, the polis gave citizens the opportunity to attempt to create a legacy that would be added to the history of Athens. By performing great acts and/or speaking well they could both honor the world and make it shine. Arendt uses the language of shining at a couple points in the text—this is language used by later Heidegger and followers of Heidegger.[19] For the Greeks, it is not enough to be another world—there are many worlds. Instead, we must be the best world and this means making our world shine through deed, speech and art. Thus, while an individual might gain prominence and recognition in the polis through great acts or great speech (which, in turn, could be commemorated through art), the shine of these actions reflects on the world itself and makes it shine also. Thus, the agonistic polis is a means of promoting excellence both in individuals and in the world itself.

My claim is that although Arendt never explicitly states it, her description of making the world shine through great action and speech is an act of love—love of a world. There are, of course, other ways to preserve and love a particular world. We could look to anthropology and history and seek to compare and contrast how different cultures and communities have preserved and loved their worlds. If we keep in mind the porousness of the premodern self, we can see that the self is intimately connected with the world. The glory of that world is the glory of the individual and their outstanding acts and speech that make that world shine.

At this point, it is tempting to turn to this analysis of Arendtian *amor mundi* directly, but there is some further work that needs to be done concerning the decline of world and the rise of modern world alienation. As I mentioned at the beginning of this chapter, Arendt never discusses *amor mundi* directly in *The Human Condition*. Instead, the text offers a genealogy of world alienation. In so doing, the text points to *amor mundi* like negative theology points to God. Without actually describing God and instead discussing what He is not, negative theology nonetheless creates a space for understanding God "negatively." Similarly, by describing the decline of world and the rise of world alienation, Arendt is pointing indirectly to the importance of world and love of a world.

In the remainder of this chapter, I will follow her "negative" approach and discuss the genealogy of world alienation. The task of attempting to outline the positive understanding of what Arendt means by *amor mundi* will be explored in the next chapter.

WORLD ALIENATION

This chapter has been focused on what Arendt means by "world." Having clarified what she means by world, in the remainder of this chapter I will discuss her genealogy of world alienation and its implications. In other words, if world was so important to premodern peoples, what happened to it?

After focusing for one chapter each on labor, work, and action, Arendt immediately launches into a discussion of world alienation. In this particular section, she mentions three historical causes of modern world alienation:

> Three great events stand at the threshold of the modern age and determine its character: the discovery of America and the ensuing exploration of the whole Earth; the Reformation, which by expropriating ecclesiastical and monastic possessions started the two fold process of individual expropriation and the accumulation of social wealth; the invention of the telescope and the development of a new science that considers the nature of the Earth from the viewpoint of the universe.[20]

These are the three modern historical developments that Arendt directly connects to world alienation. In addition to these, Arendt also discusses in detail two other factors that are a part of a very broad arc of change: first, the shift away from the politically focused *vita activa* of the ancient Greek polis to the otherworldly focus of the *vita contemplativa* that became ascendant with Plato and held sway until the rise of modernity. Second, Arendt claims that modernity turned away from the *vita contemplativa* toward the *vita activa*, but not in a way that recovered the freedom and emphasis on action that marked Greek politics, but instead in a manner that stressed labor and social control. In her account, modernity rejects the otherworldliness of Platonic and medieval thought, but turns to the creation of a labor-based society that is fundamentally unfree instead of reestablishing the freedom of the polis. But a labor-based society cannot establish a world and instead results in the fragmentation and proliferation of worlds. As such the modern emphasis on labor greatly contributes to modern world alienation.

There are at least five factors that play a role in the genealogy of world alienation. I will focus first on the ancient shift from the *vita activa* to the *vita contemplativa*, then turn to the three historical causes of modern world alienation and finish with Arendt's discussion of modernity as a labor-driven society.

FROM THE *VITA ACTIVA* TO THE *VITA CONTEMPLATIVA*

The first cause of world alienation is not a modern one, but an ancient one: the rise of the *vita contemplativa* and the otherworldly focus it entails.

The agonistic competition of the Greek polis was intensely this-worldly: it was focused on preserving and adding to the glory of this world. Plato rejected this. As we saw in chapter one, Plato was not content with what he claimed was the failure of ancient democracy. Additionally, he, like many other thinkers in the ancient world, was suspicious of the ways that the body, emotions, and love can prevent us from being rational and orderly. In his earlier thought, the contrast to the chaos of this world—both the chaos and rhetoric of the polis, but also the chaos of fallible humans driven by physical desires for food, sex, and love—is the intellectual realm, accessible through philosophic practice. In his later thought, the entire sublunary sphere is treated as the realm of becoming and change. As such, it is also the realm where evil occurs. Outside of this, in the supralunary spheres, there is order and regularity. In both Plato's earlier and later thought, this world is treated as lacking and in need to improvement through study and imitation of another realm.

Arendt claims this otherworldliness was taken over by Christians who also had a different realm to emulate: heaven. She claims that Christianity downplayed the political importance of "the world" and instead focused on kind of "philosophic *apolitia*."[21] Thus, while freedom for the Greeks entailed being free from labor and work, Christian otherworldliness rejected not only labor and work but also action and politics. This results in a general indifference to this world that she thinks marks the entire period between the decline of the polis and the rise of modernity. Instead of glory in the polis being the highest form of human activity, the Platonic and Christian *vita contemplativa* takes contemplation to be the highest form of activity. Note that this otherworldliness is both a rejection of the political world *and* the Earth in favor of the cosmos or heaven.

When Jesus says in Mt 22:21 that we should "Render therefore unto Caesar the things which are Caesar's; and unto God the things that are God's" many Christians read this as a radical separation of this fallen world from the next. For some Christians this world does not matter—what matters is heaven and God. And while it is true that after the rise of modernity with its this-worldly focus there is a corresponding this-worldly Christian protestant ethic (as documented by Weber); even today many Christians distinguish the "ways of the world" from the "ways of God." The followers of God are supposed to be different from "the world" and its fallen ways and, whether contemplatively or actively, turn to another world: heaven. Thus, when Christians actively seek to intervene in the world, they do so to change this fallen world and save it from itself, to make heaven here on Earth.

Like Nietzsche, who denigrates otherworldliness as nihilistic and life denying, Arendt is fairly disdainful of the *vita contemplativa*. Since her concern is *amor mundi*, she claims the anti-political otherworldliness of the *vita contemplativa* is one of the primary culprits in the decline of world and fundamentally opposed to political commitment and love of an Earthbound world.

Despite the rejection of the *vita contemplativa* by modernity, Arendt does not think that modernity recovers the love of world that marked the ancient polis, and as such the modern *vita activa* also fails. This claim is about modernity. It occurs in addition to the three historical developments that she links to modern world alienation. Arendt's concern with the modern *vita activa* will be clearer after we examine these three historical developments, so I will turn to these next.

The Shrinking of the Globe

The first modern historical development that contributes to the decline of world is the exploration of the world and the corresponding shrinking of the globe. Through the gradual exploration of the Earth, "the mapping of her lands and the charting of her waters . . . man has taken full possession of his mortal dwelling place and gathered the infinite horizons, which were temptingly and forbiddingly open to all previous ages, into a globe whose majestic outlines and detailed surface he knows as he knows the lines in the palm of his hand."[22] The unknown and the unmapped are mysterious. When one goes "off the map" who knows what one might find? Lost civilizations, hidden treasures, and bizarre creatures? The very openness of the unknown makes it appear and feel larger. At first, when I have lived in large cities, the city itself is mysterious and expansive. But with time, as I come to know the city, it becomes less mysterious, less open-ended and—in experiential terms—smaller.

This is aided, of course, through the "abolition of distance" and the "shrinkage of space" made possible through modern trains, planes, cars, and boats.[23] Previous to these forms of transportation the world was larger due to how long it took to travel around. People who lived in towns that are now hours away might never interact with each other before modern forms of travel. As such, they were more likely to develop their own dialect or language, to deviate in terms of traditions—in short—to be different worlds. Distance isolates and this isolation enables the development of a unique world. When this distance is overcome and a particular world comes into contact with other worlds, the power or hold of a particular world is lessened—now it is just one world among others.

Arendt argues that prior to the shrinkage of space enabled through modern transportation:

> There is the infinitely greater and more effective shrinkage which comes about through the surveying capacity of the human mind, whose use of numbers, symbols, and models can condense and scale Earthly physical distance down to the side of the human body's natural sense and understanding. Before we knew how to circle the Earth, how to circumscribe the sphere of human habitation in days and house, we had brought the globe into our living rooms to be touched by our hands and swirled before our eyes.[24]

When the Earth becomes a small object in our hands, we become distanced from it. Numbers, symbols, and models are abstractions, and they disconnect us from the Earth. The Earth ceases to be "her" and instead becomes a quantifiable and quantified "it." The result is Earth alienation. But the shrinking of the Earth does not just affect our understanding and relation to the Earth but also our understanding and relation to a particular human world. These worlds take place on the Earth, and when the Earth shrinks, so too do all the worlds on the Earth.

This exploration did not end with the Earth. It has continued on into the universe, which we have discovered is incomprehensibly vast. As we saw with the shift from the Earth-centered Platonic cosmos to the expanding universe in chapter 1, the Earth becomes very small. While the telescope enabled the discovery of the vast universe, the microscope revealed a new world of tiny organisms and matter. The death of the cosmos led to both an incomprehensible enlargement of the world around us and the discovery that our own bodies and our own psyche were themselves hypercomplex worlds. The expansion out and the expansion within made the Earth and any particular world smaller and less important.

If this analysis is correct, then the existence and preservation of a world requires, to some degree, limitation. The more worlds we know, the less likely we are to be committed to only one.

Expropriation

The second modern historical development that contributes to the decline of world is the modern destruction of medieval worlds through expropriation.

A particular cultural/political world is located in space and rooted in a particular place. According to Arendt's reading of the Greeks, to be a citizen of a particular world meant to own property in that world—that is, in that *place*.[25] Wealth and property were two different things. A foreigner, no matter how wealthy, has no say in the affairs of this world because they have no place in it, they don't belong to it. On the other hand, a particular family might have a place (own and live on a piece of land) in the world, and therefore the "free" head of the household might obtain status as a citizen and participate in the *bios politicos* even if they were not as wealthy as a foreigner. Arendt claims this connection with place was so strong that if an individual was expelled from the polis and banished from, say, Athens, the buildings that made up their home would be destroyed.[26]

Property therefore is tied to place. Now, however, we tend to think of property as "stuff that is owned by person X" including perhaps a home or land, but also their clothing, cars, and things. In the Athens being described by Arendt, in a sense, your property owns you. Since you have a piece of that place, you are a part of that world and you have responsibilities toward that world and that place. If you fail to live up to those responsibilities, it may justifiably be taken from you and given to someone who will live up to their responsibilities as a member of this place and this common world.

These responsibilities to a particular world and place limit the growth of wealth. Here, despite her criticisms of Marx, she is agreeing with his account of primitive accumulation and its relation to expropriation. For Marx, capitalism begins with the breaking apart of the medieval world—destroying the power of the guilds, breaking the moral constraints on wealth accumulation, contesting the authority of feudal lords and, above all, "the complete separation of the labourers from all property."[27] Expropriation was one of the primary means of tearing down the old order and turning peasants, who had a place in relatively "sovereign towns," into placeless proletarians forced to sell their labor in factories.[28] This process is further elaborated by Karl Polanyi, who traces how various enclosure acts in England required poor farmers and peasants to build fences around their land. Many were unable to do so and sold off the land, which was increasingly bought up by wealthier farmers or landowners who then kicked the peasants off the land.[29] This was ostensibly done in the name of efficiency, but practically it meant a complete social reorganization that left many destitute.

Arendt agrees with Marx that this was a watershed moment in the rise of modern capitalism and the project of modernity. For Marx, the breaking apart of the medieval world enables the exploitation of the proletariat. The placeless proletariat moved to large cities to sell their labor in factories. While factories made for massive increases in wealth (mostly for the few), they were built on the alienation and exploitation of the proletariat. Since the worker has no creative say over the product they are making nor any control over the finished product, they are alienated from their own labor and themselves. They are turned into machines with no dignity that must repeat mindless tasks over and over.

Expropriation, Marx tells us, is the hidden foundation of the rise of capitalism, but this brutal reality is covered over with myths about working hard:

> This primitive accumulation plays in Political Economy about the same part as original sin in theology. Adam bit the apple, and thereupon sin fell on the human race. Its origin is supposed to be explained when it is told as an anecdote of the past. In times long gone by there were two sorts of people; one, the diligent, intelligent, and, above all, frugal elite; the other, lazy rascals, spending their substance, and more, in riotous living. The legend of theological original sin tells us certainly how man came to be condemned to eat his bread in the sweat of his brow; but the history of economic original sin reveals to us that there are people to whom this is by no means essential. Never mind! Thus it came to pass that the former sort accumulated wealth, and the latter sort had at last nothing to sell except their own skins. And from this original sin dates the poverty of the great majority that, despite all its labour, has up to now nothing to sell but itself, and the wealth of the few that increases constantly although they have long ceased to work. Such insipid childishness is every day preached to us in the defence of property.[30]

Even today, we hear narratives akin to this. If you work hard you will succeed (and, of course, the ensuing fallacies that if you are wealthy you must have worked hard, and those who are not wealthy must not have done so). While Arendt may agree with much of this analysis, she thinks Marx overlooks the importance of world: "World-alienation, not self-alienation, as Marx thought, has been the hallmark of the modern age."[31] Thus not only have the industrial proletariat been torn from their world but that world itself has been torn apart.

Why must such a medieval world be destroyed? Lewis Mumford reminds us that the premodern worlds were not focused on production the way modern worlds tend to be. Instead, art, religious ritual, play, and community were the primary foci.[32] People did not exist to make stuff—they existed to play, worship, dance, act, talk, eat, and be together. These goals shaped their technology and thus new inventions tended to be less about production and more about making more beautiful art or better worshipping the gods. But there have been times historically when massive projects, like the pyramids, for example, required the creation of what Mumford calls the "megamachine"—turning humans into cogs in a huge machine to accomplish these astonishing tasks. To accomplish such tasks requires diminishing the traditional focus on play, art, ritual, and community and instead turning humans and their technical abilities into part of the megamachine.

Modernity is likely the greatest megamachine ever constructed. In order for it to succeed, it needs to prioritize the modern goals of production and efficiency over the traditional goals of play, worship, art, and community. To do so requires either destroying these traditional worlds or warping them to prioritize the goals of production and efficiency. This is precisely why, for Arendt, traditional worlds faced such a heavy attack. The responsibilities toward the world and others in a world were often moral and rooted in art, ritual, community, and play. But this gets in the way of maximizing profit.

Maximizing profit requires exploitation of the proletariat which requires their removal from their world and its destruction.

Thus while Arendt agrees that expropriation is one of the hidden foundations of modernity, her worry is with the destruction of world and the ensuing worldlessness. Without a world to organize and limit the flux of life, we instead find ourselves caught in an unending process of constant change: "a booming prosperity which . . . feeds not on the abundance of material good or on anything stable and given, but on the process of production and consumption itself."[33] This too sounds like Marx, who claims that capitalism treats capital as the end of society rather than humanity, and this initiates an endless quest for money with no stopping point. Arendt claims that we are living in a "waste economy" in which "conservation spells ruin because the very durability of conserved objects is the greatest impediment to the turnover process, whose constant gain in speed is the only constancy left wherever it has taken hold."[34]

This modern acceleration is in direct conflict to world. World stabilizes and organizes human experience. Modern capitalism destabilizes and constantly seeks to pile on more and more accessories and gadgets. This endless "process of wealth accumulation" is only possible "if the world and the very worldliness of man are sacrificed."[35]

Science and the Rejection of the Testimony of Nature

The third historical development is the rise of modern science which casts doubt on the testimony of nature. Here Arendt is pointing to a unique and new feature of modern science that distinguishes it from premodern science—that modern science often defies everyday empirical experience. This, in turn, leads to pervasive doubt about the world around us.

The key figure in this regard is Galileo. Galileo, through the use of the telescope, enabled us to see things that were previously beyond human ability to see: "that is, he put within the grasp of an Earthbound creature and its body-bound sense what had seemed forever beyond his reach, at best open to the uncertainties of speculation and imagination."[36] Other thinkers and scientists before Galileo had made various claims that were considered inoffensive by social authorities of the time because they were speculative or mathematically useful, but since Galileo made the previously invisible visible "he established a demonstrable fact where before him there were inspired speculations."[37] Reality had lied, the human senses had misled us and Galileo had used his instrument to uncover the truth. Arendt claims that the effect of this was to undermine belief in the human senses and initiate what Nietzsche called the "school of suspicion" that was later driven by Cartesian doubt and anxiety about human perception and the status of the world around us.[38]

The doubtfulness of reality furthers the displacement of humans—after all, where are we if we cannot even trust the reality of the world around us? The human individual, like Descartes, is thrown back on themselves: if the world can be doubted and the only thing I can be sure of is that I think, then am I not left with only myself? And while Descartes sought to reject solipsism by an appeal to God, the practical effect of Galileo and Descartes is a world and universe of questionable existence which could be radically different than it appears. Arendt says that this is why Heisenberg claims that now "man encounters only himself."[39]

Ancient math and ancient science were tied directly to the Earth and human experience on the Earth. Modern math and modern science have become divorced from basic human experience so thoroughly that Arendt claims nothing akin to modern math and science has existed previous to modernity.[40] Appearance is now meaningless. To truly understand requires getting behind or beyond appearance, and it is through special instruments and math that we can access the hidden reality and discover the truth. Sometimes the phenomena in question are so large or so small that they not only fall beyond the realm of the humanly visible but stretch the bounds of comprehensibility. Oddly, sometimes what is discovered thanks to math is near impossible to translate back into human experience. One of the truly strange results is that some scientific communities are speaking to each other in math and data and struggle to communicate what they are "seeing" with the rest of us. Speech becomes impossible unless one speaks that particular language.[41]

As a result of this situation, modern doubt has replaced ancient Greek wonder as the fundamental way to approach the world.[42] Furthermore, since the human senses have become so dubious, the standard for finding truth is no longer just seeing it, but reproducing it. It is not enough to see the data, one must be able to produce and reproduce the phenomena in question to be sure that one understands what is happening.[43]

This situation results in a breakdown of "common sense"—the shared understanding of the world that marked premodern experience.[44] "What men now have in common is not the world, but the structure of their minds."[45] As Descartes concluded, with the world around us becoming so doubtful what we can be sure of is that we are thinking. This opens up a whole realm of interiority and seems to call for an inward exploration and introspection. World ceases to be the center of human experience. A particular sociopolitical world has become a small part of a small planet in a vast universe. Similarly, an inward depth has opened up as the self has become its own mysterious and deep frontier. From these depths have emerged all sorts of new imaginary worlds: Hogwarts, Middle Earth, Disneyworld, the land of Ooo. The traditional world—smaller and doubtful—is now caught among all these other worlds.

Perhaps the internet represents the epitome of this inversion away from identifying with a particular world to which one is intimately connected and responsible, to instead flittering around from one fungible alternative world to the next, connected to none, trading roles as convenient, free to float away into the endless digital eternities, tied only to Earth by the inconvenience of a body.

THE MODERN VITA ACTIVA AND CONSUMERISM

The final historical development that Arendt discusses in detail is consumerism. Arendt has a unique understanding of consumerism that sees it as the result of the modern rejection of the *vita contemplativa* and return to the *vita active*—except this modern return to the *vita activa* inverts the traditional Greek valuation of action over work and labor. According to Arendt, modernity favors labor over work and action and thus treats society as an object to be managed and ordered. As a result, the modern *vita activa* is anti-political. This places it in opposition to *amor mundi*—love of a sociopolitical world—which was for the Greeks integrally related to politics. Thus, for Arendt to recover *amor mundi* is to recover both in understanding and practice the political from an anti-political consumeristic society that thoughtlessly misunderstands politics.

I am going to put off a full discussion of this development until the next chapter however, because of how closely related it is to understanding the positive sense of what Arendt means by *amor mundi*. Before turning to this, let me summarize the discussion in this chapter of what Arendt means by world.

SUMMARY: WHAT IS WORLD AND WHAT HAPPENED TO IT?

A particular socio/cultural/political world is located in place and has a physical presence through the buildings, roads, temples, states, and so on that humans have built through work. This physical world is shaped by the spiritual and cultural meanings and history that are the other part of such a world. The physical there-ness of a world has a reciprocal relationship with the spiritual, cultural, and historical "soul" of that world: the physical world commemorates and embodies this "soul" and the spiritual, cultural, and historical meanings suffuse the physical world with life and rich significance.

Any such world transcends the individual: she is born into a world that precedes her and will (most likely) outlast her. A world organizes experience, gives it sense, meaning, and depth. As such a world merits respect, and this may very well imply duties to it. Since the world gives us life and gives us meaning, we can feel gratitude and love toward it.

It is difficult for moderns to understand the depth of world, much like we struggle to understand the premodern experience of the cosmos. As I discussed in chapter 1, the premodern self was "porous" and open to the world and cosmos, while the modern self has become "buffered" and insulated. We do not, or perhaps only very rarely, experience a level of connection to the world or cosmos that would have been commonplace previous to modern world alienation and the death of the cosmos. As such it is difficult for us to comprehend what it would mean to truly love a world.

Moderns do experience world, but we tend to experiences "worlds" and not world. Like a museum-goer who moves from room to room which is organized in terms of "worlds" (the Aztec room, the Maya room, French surrealism, Baroque, and the like), modern experience is accustomed to multiple worlds that overlap and proliferate. We recognize that a particular world makes claim on us in terms of how we should behave, has particular meanings that are unique to that world, and demands certain forms of speech. Modern technologies have shrunk the Earth and brought what were previously different worlds into contact with each other. Furthermore, modern science has undermined the "obviousness" of the world around us. As such the world(s) for moderns has/have become more superficial, less meaningful, less compelling, interchangeable, doubtful and overlapping. While we may still commit to a particular world, we are less likely to see such a world as *the* world. As such, we feel less responsibility to the world, less gratitude to it and are less likely to experience a deep, meaningful, and lasting love toward it.

Is this so bad? On the one hand, there is immense freedom in being able to move between worlds and not be limited to a particular world. Furthermore, many traditional worlds had aspects that are worrying—tribalism, sexism, xenophobia, and perhaps other aspects that we might find unethical and disturbing. Is Arendt's concern with world backward-looking?

For Arendt, there are some fundamental reasons to lament world alienation. In positive terms, love and commitment to a world is necessary for a functioning democracy and public happiness. In negative terms, modern world alienation loses the stability and order provided by a world and has decentered and fragmented human experience in a way that can lead to nihilism. To understand this claim requires paying attention to how central world was to human life, how deeply the premodern individual was connected and identified with a world. In a sense, world was everything. Morality, meaning, and order were all tied to world. Accustomed to living in a fragmented world, moderns have a hard time imagining what it would be like to live in a world that functioned as a whole. Lewis Mumford alerts us to this difference when speaking of the implications of Galileo's discoveries:

> Galileo committed a crime far graver than any the dignitaries of the Church accused him of; for his real crime was that of trading the totality of human experience, not merely the accumulated dogmas and doctrines of the Church, for the minute portion which can be observed within a limited time-span and interpreted in terms of mass and motion, while denying importance to the unmediated realities of human experience, from which science itself is only a refined ideological derivative. When Galileo divided experienced reality into two spheres, a subjective sphere, which he chose to exclude from science, and an objective sphere, freed theoretically from man's visible presence, but known through rigorous mathematical analysis, he was dismissing as unsubstantiated and unreal the cultural accretions of meaning that had made mathematics—itself a purely subjective distillation—possible.[46]

Mumford's claim here is that Galileo cut the "totality of human experience" into at least two parts: the subjective part and the objective part. This complaint was one that was shared by Arendt and her teachers: that modern science, in its efforts to isolate "objective" reality, tore reality apart—on the one side the "objective" material reality and on the other the "subjective" aspects of life, including meaning and morality. The problem for Arendt and other thinkers like Heidegger and Husserl is that this turns morality into "values"—that is, subjective moral preferences that are in the mind of the subject. Questions of morality and meaning are divorced from reality and moved into the minds of humans. There are two problems with this. First, it undermines the most important questions of human existence by making them akin to opinion. Second, it decrees these questions unavailable to science, the most prestigious form of modern knowledge. As a result, these fundamental questions become second-tier concerns, reduced to a questionable ontological and epistemological status and torn from the "real." Stripped of their status and authority, morality and meaning seem to be up for grabs, easily discarded as "subjective" or just opinion. The ensuing situation may result in kind of moral confusion that Nietzsche described as nihilism. For Arendt and Heidegger, modern "values" are like clothes—we wear the ones we like when we want. Meaning, mattering, and morality have become flimsy and ephemeral.

For the premodern world-dweller, morality is not subjective. Meaning and morality are not ephemeral at all, they are objective in the sense that they are "out there"—whether in the cosmos or in the world. This is not to say that there were not competing moral perspectives within a world, but that morality and meaning had an objective reality, an undeniable weight.

The fragmentation of world alienation brings into doubt whether questions of meaning, mattering, and morality can be anything more than subjective preferences or the quirks of particular worlds. For Arendt, it is doubtful whether a free democracy and the ensuing public happiness that would result are possible in conditions of world alienation. To clarify this concern, let's attempt in the next chapter to outline the positive image of *amor mundi.*

NOTES

1. Young, Julian. *German Philosophy in the Twentieth Century* (New York: Routledge, 2018).
2. Taminaux, Jacques. *The Thracian Maid and the Professional Thinker: Arendt and Heidegger* (Albany: SUNY, 1997).
3. Young-Bruehl, Elizabeth. *Hannah Arendt: For the Love of the World* (Haverford, CN: Yale University Press, 2004).
4. Bernauer, J.W. *Amor Mundi: Exploration in the Faith and Thought of Hannah Arendt* (Springer, 1987).
5. Arendt, Hannah. *The Human Condition* (Chicago: University of Chicago Press, 1998), 1.
6. Arendt, *The Human Condition.*
7. Ibid., 2.
8. Ibid., 2–3.
9. Ibid., 3.
10. Ibid., 6.
11. Ibid., 52.
12. Ibid., 9.
13. Arendt, Hannah. *Between Past and Future* (New York: Penguin Classics, 2006), 186.

14. Arendt, *The Human Condition*, 32.
15. Ibid., 37.
16. Ibid., 24.
17. Ibid., 41.
18. Ibid., 197.
19. Dreyfus, Hubert and Sean Dorrance Kelly. *All Things Shining: Reading the Western Classics to Find Meaning in a Secular Age* (New York: Free Press, 2011).
20. Arendt, *The Human Condition*, 248.
21. Ibid., 14.
22. Ibid., 250.
23. Ibid., 250.
24. Ibid., 250–251.
25. Ibid., 61.
26. Ibid., 62.
27. Tucker, ed., *The Marx and Engels Reader (Second Edition)* (New York: W.W. Norton & Company, 1978), 432.
28. Tucker, *The Marx and Engels Reader*, 433.
29. Polanyi, Karl. *The Great Transformation: The Political and Economic Origins of Our Time* (Boston: Beacon Press, 2001).
30. Tucker, *The Marx and Engels Reader*, 431–432.
31. Arendt, *The Human Condition*, 254.
32. Mumford, Lewis. "Tools Users vs Homo Sapiens and the Megamachine" in Robert C. Scharff and Cal Dusek, eds. *Philosophy of Technology* (Malden, MA: Wiley Blackwell, 2014), 381-388.
33. Arendt, *The Human Condition*, 253.
34. Ibid., 252–253.
35. Ibid., 256.
36. Ibid., 260.
37. Ibid., 260.
38. Ibid., 260.
39. Ibid., 261.
40. Ibid., 264.
41. Ibid., 4.
42. Ibid., 273.
43. Ibid., 274.
44. Ibid., 280.
45. Ibid., 283.
46. Mumford, Lewis. *The Myth of the Machine: The Pentagon of Power* (New York: Harcourt Brace Jovanovich, Inc., 1970), 57.

Chapter 4

Arendt and *Amor Mundi*

What Does It Mean to Love *the World?*

In the last chapter, I discussed what Arendt means by world and discussed her genealogy of world alienation. According to Arendt modern world alienation and world fragmentation make it difficult for us to understand what is meant by world, much less what it would be like to be committed to and love a particular world. I also pointed out that despite her original intention to use the title *Amor Mundi*, the topic of love is actually quite sparse in the text. Why, if the text was supposed to be about love of the world, is there so little a discussion of it?

I claimed that the genealogy of world in *The Human Condition* functions like a negative theology—it points to the phenomenon of love of the world by describing a void left with the modern loss of world. In this chapter, I will construct or reconstruct a positive image of Arendtian *amor mundi*. Ultimately, I am seeking to articulate three different kinds of *amor mundi*: love of the cosmos, love of a world, and love of the Earth/Nature/place. The first two chapters are relating with love of the cosmos, this chapter will finish articulating love of a world, and the next chapter will discuss love of the Earth/Nature/place.

THE AGONISTIC POLIS

Is the agonistic polis emblematic of *amor mundi*? Arendt uses the agonistic polis of Athens as her primary example of world. As we saw in the last chapter, world refers both to the physical objects created by humans that make up a particular place including buildings, roads, statues, temples, and so on, and the spiritual and social traditions that make sense of human experience. A particular world transcends each individual who is born into it and will likely exist long after them. If we keep in mind the porousness of the premodern self, then we could say that each individual is so intimately tied to this world that they cannot be separated from it.

Athens is a particularly compelling example of world and commitment to world for Arendt. On her reading, Athens (the world of Athens) demanded respect and honor. She claims that the Athenians organized themselves in a way that maximized the possibility for citizens to do great acts and give great speeches. In so doing, they could add to the glory of Athens. The result was an agonistic competition to stand out and create a legacy that would bring glory to oneself within this tradition, but also add to it and make it shine.

While Arendt does not explicitly use this language, this deep connection and commitment to this world is a kind of *amor mundi*. The citizens of Greece were proud of their world, loved it, and likely felt loved by it. Barbarians had no such accomplishment, in their view. Athens was the height of civilization.

Does Arendt's use of Athens as an example of world and *amor mundi* mean that she thinks it is something that we should emulate? Clearly, some readers of Arendt have found her description of the ancient polis quite compelling. Chantal Mouffe claims that "Arendt is often considered the representative of agonism" in contemporary debates about "the best way to envisage democratic politics."[1] For those who feel this way, Arendt's use of the polis is not merely an example of world, but an example of a good world that we should emulate.

Here agonism is held up as the alternative to models like Rawls and Habermas that stress democratic deliberation. The Rawlsian and Habermasian models draw inspiration from classical liberalism and see democracy as a matter of careful, rational debate among people with different ideas who remain committed to working through these differences together. The claim in the democratic deliberation model is that rational consensus will win the day if carried out responsibly. Agonism, on the other hand, stresses the way that politics is competitive and full of strife. Mouffe holds up Bonnie Honig's *Political Theory and the Displacement of Politics* as an example of a feminist use of Arendt to defend the need for contestation in democracy. Mouffe herself approaches agonistic democracy from a more Marxist position that stresses not just agonism but antagonism. While Mouffe claims that Arendt is the key figure in agonistic theories of democracies, she herself thinks Arendt fails "to acknowledge the hegemonic nature of every form of consensus in politics or the ineradicable character of antagonism."[2]

Interestingly, while Arendt was an influence on agonistic theories of democracy, she was also an influence on the deliberative model of Habermas. His *The Structural Transformation of the Public Sphere* references *The Human Condition* and was published in 1962, one year before she published *On Revolution*—a work that has some remarkable parallels to it. Both

The Structural Transformation of the Public Sphere and *On Revolution* are attempting to think through moments of intense public debate whether in the salons or during revolution. Both point to historical moments that revivify the intense political focus on world that Arendt describes in the Athenian polis. But despite his background of Marxist critical theory, Habermas does not stress agonism the way that Mouffe or Honig do. *The Theory of Communicative Rationality* explicitly contrasts "communicative rationality" with "strategic action"—if anything Habermas presents strategic action as something in need of being overcome through communicative rationality.

Thus, some thinkers hold Arendt up as a foundational proponent of agonistic democracy, while others who were also inspired by her do not emphasize this agonism. There are good reasons for claiming that Arendt is a key thinker for both models. In terms of *amor mundi*, however, I do not think we have to choose between these options. In other words, there can be multiple models of *amor mundi* in the Arendtian sense (i.e., multiple examples of love of a world). If this is the case, then the democratic polis of ancient Athens could be described as a model of agonistic *amor mundi*. Arendt also offers a nonagonistic model of Greek *amor mundi*: the Socratic model of dialogical midwifery and mutual improvement of opinion. The Socratic model was never fully adopted in Western thought; instead, it gave preference to Plato's truth-based model. However, I want to examine Arendt's discussion of the Socratic model and its implications.

SOCRATES, OPINION, AND FRIENDSHIP

Arendt opens an essay entitled Philosophy and Politics with the following claim: "Our tradition of political thought began when the death of Socrates made Plato despair of polis life and, at the same time, doubt certain fundamentals of Socrates' teachings."[3]

Specifically, Arendt says, what Plato doubted was the "validity of *persuasion*."[4] Persuasion was of fundamental importance in the Greek polis and democracy, since persuasion is the way free people must try to convince each other. Recall from the last chapter that the polis was a place of freedom where there were no bosses or hierarchy allowed. To be free meant to be free from necessity. This freedom is lost if violence or force are exercised to bring about a certain result. Only by discussion and debate that respected the freedom of others, could the polis remain a space of freedom.

For Arendt, the interaction of free individuals *is* politics. Plato's disillusionment with the polis and democracy is, on her reading, a rejection of action and politics in favor of philosophy:

> Closely connected with his doubt about the validity of persuasion is Plato's furious denunciation of *doxa*, opinion, which not only ran like a red thread through his political works but became one of the cornerstones of his concept of truth. Platonic truth, even when *doxa* is not mentioned, is always understood as the opposite of opinion. The spectacle of Socrates submitting his own *doxa* to the irresponsible opinions of the Athenians, and being outvoted by a majority, made Plato despise opinions and yearn for absolute standards. Such standards, by which human deeds could be judged and human thought could achieve some measure of reliability, from then on became the primary impulse of his political philosophy . . . that is, to introduce absolute standards into the realm of human affairs, where, without such transcending standards, everything remains relative.[5]

Here politics and epistemology are working together. Plato wants the truth as a solution to a political problem: the unpredictability and unreliability of the polis. Put in the terminology of *The Human Condition*, Plato wants to eliminate action. This is nothing short of a coup—Plato is seeking to overthrow the freedom of the polis and replace it with the truth of the philosopher. This is why Arendt claims what Plato is up to amounts to establishing "the tyranny of truth."[6] Supposedly, the philosopher has access to the truth, which is not mere opinion, but something that is eternally true. By leaving the cave and ascending to the realm of forms or ideas, the philosopher transcends the world of mere opinion and returns to it with truth.

But the problem is that those that remain in the cave may not believe the philosopher who has ascended into higher realms. The truth appears as another opinion when brought to the polis. Plato, now a philosopher uninterested in action and free politics, wants to cut through the noise of the polis and impose the truth. Arendt claims that Plato's invocation of "the Hereafter" and the allegory of the cave are meant to "frighten rather than merely persuade the audience."[7] Since the citizens of the polis were too stupid to see the truth of Socrates' arguments, they must now face the wrath and "violence" of the truth.[8]

In so doing, Plato is not only rejecting the polis and democracy but also the philosophical approach of Socrates. According to Arendt, Socrates' philosophical approach was much more akin to the persuasion of the polis than the truth of Plato:

> To Socrates, as to his fellow citizens, *doxa* was the formulation in speech of what *dokei moi*, that is, of what appears to me. This *doxa* has as its topic not what Aristotle called the *eikos*, the probable, the many *versmilia* (as distinguished from the *unum verum*, the one truth, on the one hand, and the limitless falsehoods on the other), but comprehended the world as it opens itself to me. It was not, therefore, subjective fantasy and arbitrariness, but also not something

> absolute and valid for all. The assumption was that the world opens up differently to every man, according to his position in it.[9]

Let me unpack this complex passage. Each person's opinion is based on how the world appears to them. The same world appears to all of us and yet we each have a certain perspective or understanding of it. No one individual has *the* correct, privileged view of the world. But this does not mean that all perspectives are as correct or as good as all others. Since we all have limited perspectives, we can better understand the world by sharing our perspectives with each other. The famous Indian analogy of the blind men touching the elephant stresses that no one individual has all the answers. The man touching the elephant's leg thinks it is a pillar, the man touching the ear thinkers it is a tent, and so on. Only by coming together as sharing their perspectives can they understand what they are touching. But it is also more complicated than merely sharing perspectives, because some perspectives might be confused or mistaken. As such, the task is not just to hear other perspectives, but to work through those perspectives and improve them. This is what Socrates himself called maieutic, the art of midwifery: "he wanted to help others give birth to what they themselves thought anyhow, to find the truth of their doxa."[10] Notice that Socrates is not telling others what to think, but helping them better understand their own perspective or opinion. If all members of the polis seek to improve their opinions the entire polis would be enriched, because they would get to see more perspectives and understand more sides of what they are all seeing and experiencing. On this model, the role of the philosopher is not to rule with truth, but be the gadfly that seeks to help citizens improve their opinions.[11]

On Arendt's reading, Socrates is not rejecting the logic of the polis, but trying to improve it. He does not abandon action, freedom, and politics, but seeks to make the agonism of the polis less antagonistic and more friendly: indeed, Arendt claims that "politically speaking, Socrates tried to make friends out of Athens's citizenry."[12] Instead of trying to best each other, "they become equal partners in a common world—that they together constitute a community. Community is what friendship achieves."[13] To mutually support each other is to enrich our common world:

> This kind of understanding—seeing the world . . . from the other fellow's point of view—is the political kind of insight *par excellance*. If we wanted to define, traditionally, the one outstanding virtue of the statesman, we could say that it consists in understanding the greatest possible number and variety of realities . . . and, at the same time, in being able to communicate between the citizens and their opinions so that the common-ness of this world becomes apparent. . . . Socrates seems to have believed that the political function of the philosopher

> was to help establish this kind of common world, built on the understanding of friendship, in which no rulership is needed.[14]

It should be clear that this is another type of *amor mundi*: instead of the agonistic love of the world of the polis, Socrates is suggesting a version that shares some of its fundamental commitment, but emphasizes friendship, dialogue, and mutual improvement through sharing and maieutic refinement of perspectives and opinions. For Arendt, Socrates' friendly alternative to the agonistic polis is important because it retains the commitment and love of a common world, the appreciation for the plurality of humans, and the respect for democracy that were so key to the polis, while potentially finding a way to overcome the agonism that Arendt claims eventually tore the polis apart. She claims the "agonal spirit . . . eventually was to bring the Greek city states to ruin because it made alliances between them well-nigh impossible and poisoned the domestic life of the citizens with envy and mutual hatred."[15]

In light of the Socratic alternative, although the agonistic polis may indeed be an example of a kind of *amor mundi*, it is not the one that Arendt prefers. If this is the case, it does not lessen the importance of the agonistic model held up in *The Human Condition*. There it is being used as a foil to our contemporary understanding of politics, democracy, and world. Arendt's description of the agonistic polis is a major influence on contemporary theorists of agonistic democracy. However, if she actually prefers the Socratic friendship model, then perhaps Habermas is closer to the spirit of *amor mundi* that Arendt seeks.

This creates some problems for us, however, because the Socratic model was never implemented. After Socrates' death, Plato suggested an alternative model of philosophy that rejects the fundamental commitments that Socrates' friendship model shared with the democratic polis: the commitment to democracy, action, and plurality. Plato instead wants the tyranny of truth and the rule of the philosophers. According to Arendt, Western thought has embraced the Platonic approach and long forgotten both the democratic polis and the Socratic alternative to it. As such the Socratic friendship model of *amor mundi* is a long-lost alternative that she is seeking to unearth—one of the pearls she has recovered while diving.

THE TYRANNY OF TRUTH AND MODERN CONSUMERISM

Western thought's preference for Plato over Socrates has radical implications for our contemporary understanding of epistemology and politics. Perhaps surprisingly, Arendt claims the modern tyranny of truth has led to consumerism. As discussed in the last chapter, Arendt traces five historical shifts that led to modern world alienation. The first is the ancient turn away from the democratic polis toward contemplation that Arendt claims was initiated by Plato. This led to a period of otherworldly focus on heaven, God, and/or the cosmos. The next three shifts that contributed to modern world alienation are all modern: the shrinking of the globe (which decenters us from a political world), the expropriation of property (which literally tore apart the medieval worlds), and the rise of modern science which led to a distrust of the testimony of nature. The fifth feature is decisive, but I left it for this chapter because it is best understood in light of the tyranny of truth proposed by early Plato. We are now in a position to better understand the fifth feature of modern world alienation: modern consumerism.

Arendt has a unique and powerful understanding of the rise of modern consumerism and it ties back to Plato's dream of the tyranny of truth. It is interesting to note that Arendt thinks that Plato's anti-democratic dream of radical social reorganization in light of truth was not the original aspect of his philosophy that became dominant. Rather, it was the anti-democratic otherworldly focus of his later philosophy that became dominant in antiquity. Recall that in chapter 1, we saw Nussbaum's claim that there is a marked difference between early Plato and later Plato. Arendt and Nussbaum agree here: early Plato was interested in radically remaking society while later Plato is interested in contemplating the cosmos. There are connections between later and earlier Plato, for example, human life could indeed be restructured to emulate the cosmos according to later Plato. But Nussbaum and Arendt think later Plato turns his gaze to the heavens and emphasizes contemplation in a way that is quite different from the radical plans for social reorganization that seem to motivate earlier Plato. It is with modernity that we see a rejection of the *vita contemplativa* of later Plato for the aggressive, interventionist *vita activa* of early Plato. Arendt is clear that the modern return to the *vita activa* does not return to the agonistic polis of ancient Greece or the friendly *amor mundi* of Socrates, rather modernity turns to the tyranny of truth of early Plato.

Plato, in seeking to overcome the chaos and unpredictability of the plurality and action of the polis, turns completely away from the Greek understanding of politics as freedom of citizens in the *bios politikos*—the second life that was enabled by the labor and work done in the private sphere. Recall that the private sphere was the realm of necessity. The family and the private sphere in which it functioned in ancient Greece were not free.[16] The repetitive tasks of labor occurred in the home: making food, cleaning, reproduction, childrearing, and so on. The private sphere was a managed sphere where the physical needs of everyone involved took priority. The original meaning of "economics" was tied to housekeeping or management of the home.[17] We can see this with the high school class of home economics in which students (traditionally mostly female) learn skills such as cooking, sewing, financial management, and child development. These classes are often referred to as "Home Ec," but what may be forgotten in this abbreviation is that the "ec" in "home ec" refers to economics. Only the head of the household and perhaps some of his male children were allowed to exit this managed and controlled space of home economics and allowed to enter the polis where freedom occurs.

When Plato rejects the polis and the principles of freedom and action that ruled the polis, Arendt claims he is seeking to impose the logic of the private sphere on the polis. In other words, she claims that Plato wants to run the entire city like a giant "super-human family" with the philosophers in charge.[18] Arendt claims that while Plato imagined what it would be like do so in the Republic, he never carried this out in practice. When modernity rejected the *vita contemplative*, it did not seek to reinstate the logic of the polis but instead attempted to carry the Platonic project of remaking society as a giant, highly managed super family. Arendt refers to this super family as "society."[19]

The rise of society then sees all members of a city or later a nation as part of a giant family that is in need of proper organization and direction. But what exactly is the primary concern or direction of modern society? This question is very complex since there are multiple competing goals in modernity. In the last chapter for example, we saw that one of the goals of modernity was to break apart the medieval world in order to allow for the accumulation of capital. Medieval traditions and morality impeded the sort of exploitation and social organization that make modern capitalism possible. Expropriation tore communities apart and allowed for the creation of the industrial proletariat. Thinkers such as Marx, Foucault, and Castro-Gómez have critiqued this dark side of modernity.[20]

Arendt argues that even the ostensibly more benevolent goals of modernity have radical consequences that are sometimes not apparent at first. In *On Revolution*, she claims that one of the motivations of seeking to reorganize modern society is to assure the well-being of all in society. This goal focuses on the physical well-being, that is, biological survival, of the family and includes new (modern) concerns with poverty and happiness. Arendt claims that poverty was so ubiquitous in the premodern world that it wasn't seen as a problem to be solved, but as the inevitable state of affairs of any city. With the modern rise of society, all the people in a city or nation were seen are part of a giant family and poverty appeared as a social problem (and not an inevitability). Arendt points out that thinkers like Marx saw poverty as the result of the way a particular society was organized—that some were exploiting others—and sought to reorganize society in a way that would eliminate poverty.[21] Critically for her argument, this intense focus on the biological well-being of everyone in society places the emphasis on labor and not work or action. According to Arendt, modern economics and the rise of society completely inverts the traditional order of the Greek city-state which valued action (and freedom) over work and labor and instead places labor over work and action. To be clear, this essentially eliminates the Greek emphasis on freedom, the second life of the bios *politikos*, the polis and action. The (unfree) logic of the private sphere is writ large over all of society and structurally society no longer provides a specific and articulated place (a polis) for free action. Thus, Arendt makes a rather stunning assertion: "The politically most pernicious doctrine of the modern age [is] that life is the highest good, and that the life process of society is the very centre of human endeavor. . . . Not freedom but abundance became the new aim of revolution."[22]

Arendt is not saying that the biological life of humans (suffering, hunger) does not matter. But this modern focus on life and labor (which feeds biological life) leaves out the questions of *why* life matters. It supports bare life with no clear indication of what life should be—except the vague goal of happiness. According to Arendt, the ancient Greeks were not interested in bare life, but in freedom, glory, and *amor mundi*. The purpose of life was not just to be alive, nor to be happy, but to win glory in the polis and honor Athens.

The problem for Arendt is that if you make labor the primary concern in society instead of work or action, then everything will center on labor—which is potentially endless. The "daily needs and wants" end up becoming a new "taskmaster," and we become enslaved to labor.[23] In *The Human Condition*, Arendt points out that the strange result of the modern emphasis on labor is that we don't know what else we would do besides labor. She points to the phenomenon of automation—robots replacing human labor—and points out that freedom from labor was precisely the goal in many premodern cultures. But robots taking human jobs has become a curse in modern society: If the robots do all the work, what will we do with ourselves?

> The modern age has carried with it a theoretical glorification of labor and has resulted in the factual transformation of the whole of society into a laboring society. . . . It is a society of laborers which is about to be liberated from the fetters of labor, and this society does no longer know of those higher and more meaningful activities for the sake of which this freedom would deserve to be won. . . . What we are confronted with is the prospect of a society of laborers without labor, that is, without the only activity that is left to them.[24]

This really is a bizarre outcome. What are we to do if there is no more need to labor? For the Athenians, you got to the polis, be free and seek glory. For Socrates, you dialogue with others as part of a common world. For later Plato, you contemplate the cosmos. While the modern inversion of labor over work and action makes sense from the perspective of poverty and social well-being, Arendt claims it has the disturbing effect of undermining world and freedom as goals of society.

When action is no longer a goal of society in modernity, the institutional structures that existed in Athens—including the physical space set aside as the polis—are not incorporated into the modern world(s). When work is no longer a goal of society in modernity, we no longer tend to build the physical world with the pride and meaning that marks many premodern worlds. In terms of the quality of the world around us, if labor replaces work, then the things we make tend to be more like the products of labor, which are temporary and meant to be consumed, than the products of work, which are meant to be permanent or semipermanent and become a part of a world. As a result, things like tables, chairs, pots, houses, and so on, turn into cheap-to-make, mass producible objects that are not intended to last long.

Even objects that are supposed to be permanent are increasingly made as if they are temporary. Our language reflects this: we call the things we make "consumable" goods instead of long-lasting products and humans become "consumers" whose primary action in life is to consume objects. Speaking of the quality of things in a consumer society Arendt says:

> Their very abundance transforms them into consumer goods. The endlessness of the laboring process is guaranteed by the ever-recurring needs of consumption; the endlessness of production can be assured only if its products lose their use character and become more and more objects of consumption, or if, to put it in another way, the rate of use is so tremendously accelerated that the objective difference between use and consumption, between the relative durability of use objects and the swift coming and going of consumer goods, dwindles to insignificance.
>
> In our need for more and more rapid replacement of the worldly things around us, we can no longer afford to use them, to respect and preserve their inherent durability; we must consume, devour, as it were, our houses and furniture and cars as though they were the "good things" of nature which spoil uselessly if they are not drawn swiftly into the never-ending cycle of man's metabolism with nature.[25]

If labor replaces work, and we think of the objects in our world as consumer goods, then we lose the sense of belonging to a stable world. This stability is threated because we are constantly making more and more stuff. Increasingly, we own very few objects that are meant to be passed on from parents to children. On the contrary, the things we buy tend to be cheap. Since we are making more stuff and the stuff we are making tends to be flimsy, we don't hold onto it. We buy more and more, but we also throw out more and more since it is not made to last. This means, of course, that we create massive amounts of litter and pollution. Importantly, as the world of lasting things that mattered so much in premodernity is devalued in favor of labor and mass produced stuff, we have fewer special places and things. The world becomes less sacred and more fungible.

For Arendt, this is potentially a disaster for world and Earth. World becomes devalued and consumable: we voyeuristically tour through different worlds and revel in fantasy worlds and alternative realities. These worlds are worn like clothes: in the morning this, in the evening that. Environmentally, the sheer amount of trash we produce piles up behind us as we consume. But Arendt is also concerned about the implications of this for the thinking capacities of modern humans. Labor is focused on activities that lend themselves to repetition and automatic processes: no matter how much food you prepare, you will always need to prepare more later. Modern consumerism encourages us to keep making more and more. But this means that much of what happens in our society is continually subjected to automated processes that never stop. There is a mindlessness to automated processes: they just go on forever. If you have ever had a job in a factory-like setting, where you do the same process over and over, you know it can be mind-numbing. Also, independent and critical thought are often discouraged in a setting of repetitive labor activity. It is not an accident that humans are often replaced by robots and consumers are compared to zombies in modern film.

When labor is valued so highly and everything is submitted to its logic, then *thinking is often systematically discouraged.* Notice the important consequence: Arendt says we *think* less in a society focused on automation and efficiency. We make more and more, but we don't think about whether this is good or not. We don't teach people to think, rather we prepare them for jobs. Furthermore, we increasingly conceptualize thinking itself as solving problems, making things more efficient, increasing productivity, and so on: in other words, we greatly value what Arendt calls *cognition*, but NOT *wonder*, *contemplation*, and forms of thinking that are "useless."[26] And, yet, these are the kinds of thinking that are needed to question the social order and critique the way we are doing things.

The result of this modern focus on labor is not just the loss of world and spoiling of the Earth, but a worrisome pervasive cultural thoughtlessness: "Thoughtlessness—the heedless recklessness or hopeless confusion or complacent repetition of 'truths' which have become trivial and empty—seems to me among the outstanding characteristics of our time."[27] This thoughtlessness prevents us from seeing our situation. It erases alternatives, and it has erased both world and Earth.

The situation is particularly dire then, not just because we are consuming the world and destroying the Earth, but because we may be losing the capacity to think in a way that would allow us to recognize and engage with our predicament. This is why Arendt claims we need "to think what we are doing" and that this is the "central theme" of *The Human Condition*.[28] Read in terms of thinking, *The Human Condition* is attempting to show us how the glorification of labor in modern society results in a pervasive thoughtlessness about our own practices. Everything has become so automated and driven by efficiency, that we rarely take the time to stop and think. More than this, the forms of thinking that we encourage (cognition) are not the kind that are very good at questioning the way we do things (wonder, contemplation). I have discussed this problem at some length in my previous book *How the Neoliberalization of Academia Leads to Thoughtlessness: Arendt on the Modern University*. In that book, I offer an analysis of this thoughtlessness and how it is manifest in contemporary academia.

If the task in my previous book was to think through our modern thoughtlessness, this work is seeking to recover the lost treasure of *amor mundi*. If we want to learn to love a world, we will have to reject the consumerism of modern society. For Arendt, this consumerism is the outcome of emphasizing the priority of labor and well-being over building a stable world or creating spaces for freedom. Put alternatively, we need to recognize that there is much more that is important than the economy.

For Arendt, the polis is emblematic of the kinds of spaces we need. While the ancient polis was lost due to the agonism it created, she thinks Socrates has articulated a model of dialogical love of the world. This Socratic alternative stands at the beginning of philosophy as a lost alternative: the alternative that Plato and Western thought failed to follow. Despite this failure, she claims that in times of revolution, we find a pattern of moments where truly political spaces open up—in other words, we find humans working together to take responsibility for their world. These moments of revolutionary action are the topic of her late and much-neglected *On Revolution*. While Western thought has followed Plato and missed the opportunity for freedom in Socrates' dialogical model of world love, there have been moments in Western history where this possibility has resurfaced briefly. I want to follow Arendt in seeking to recognize and appreciate these lost treasures.

REVOLUTION AND THE LOST TREASURE

On Revolution is a comparison of the French and American revolutions. Like many of Arendt's works, it is misleading if taken at face value, and the philosophical importance of the text may not be immediately apparent if it is mistaken to be a work of comparative political history. Read alongside works like *The Human Condition* with an eye to the problem of *amor mundi*, *On Revolution* appears as a text that is attempting to trace moments when *amor mundi* has resurfaced—moments of true politics, moments of the return of the freedom, and joy of the polis.

The details of Arendt's account of revolution clearly hearken back to the terminology and concerns of *The Human Condition*. Immediately, in the first chapter, Arendt is considering "the meaning of revolution" in light of the modern rise of the "social."[29] She distinguished between the revolutionary goals of "liberation" and "freedom." Liberation concerns freeing individuals from forces that seek to control them, such as a tyrannical government or kings, but also, eventually, poverty. Not surprisingly, in light of the modern tendency to see "society" as a big family that needs to be properly managed, she claims that the "social question" came to be focused on alleviating poverty and improving the welfare of members of this giant family.[30] Freedom, on the other hand, is less about being liberated from something, but being free to engage in politics with others. For this to occur, we need "to build a new house where freedom can dwell."[31]

She claims that the American Revolution in particular was torn between the option of emphasizing liberty from Britain, on the one hand, and freedom of citizens to meet and engage in politics on the other. She thinks that the many small towns of America gave many Americans the opportunity to directly participate in politics—to meet with others and participate in thinking and deciding on what the community should do and be. These moments of freedom were also particularly prominent during the American Revolution when those involved needed to articulate and think through alternatives to British rule, how local communities would respond to the various situations of the revolution, what the United States should look like if it were to become independent, how the revolution should proceed, and so on. According to Arendt, these were uniquely democratic moments that involved intense public debate and community action. As such, she thinks that American political thinkers, Jefferson in particular, were aware to a larger degree that we tend to be now, of the dignity of this kind of freedom and were aware of the importance of creating spaces to foster and allow for it.[32]

How so? First, Arendt claims that the revolution had given many Americans a taste of what it meant to care for and take responsibility for a world. This was in addition to the experience they would have had with living in smaller communities with more opportunity for political involvement. She claims that there was a recognition of this freedom and an understanding that to keep this freedom alive required creating spaces for it. But there was a tension between the economic goals of accumulating and defending property and the political goals of fostering spaces for freedom and democracy. Arendt distinguishes between "private welfare" or "private happiness" and "public happiness."[33] "Public happiness" refers to the joy of being together with others and participating in the freedom of the polis. Indeed, Arendt claims that this was the "New world" version of the "Old World . . . dream of public freedom."[34] "Private happiness," on the other hand, refers to personal economic gain.

There was, then, a tension in how "happiness" was understood. One option was the "public happiness" of politics, which had appeared in ancient Greece in the polis. The other was the pleasure ("private happiness") of economic success. These are very different goals and, as Arendt claimed in *The Human Condition*, they cannot mix. The ancient polis banned labor and work because if they were allowed into the polis they would compromise the possibility of action. During and shortly after the American Revolution, there was a moment where Americans had the chance to "build a new house where freedom can dwell"—when something akin to the polis and the *amor mundi* that existed in it could have been reestablished and supported. Arendt thinks this possibility is evident in this ambivalence toward what is understood by "happiness." "For the American Revolution, it was a question of whether the new government was to constitute a realm of its own for the 'public happiness' of its citizens, or whether it had been devised solely to serve and ensure their pursuit of private happiness more effectively than had the old regime."[35] In the end, the economic won out over the political.

Once the economic is taken as the fundamental concern, the mentality of treating society as a giant family takes over, and it becomes the task of government to enable commerce, protect property, and create conditions that allow for the alleviation of poverty (whether actively intervening in society to fix poverty or creating a situation in which hard work and success are enabled and protected). As a result, "government has degenerated into mere administration, the public realm has vanished; there is no space either for seeing or being seen in action. . . . Jefferson's pride of being a 'participator in government,' political matters are those that are dictated by necessity to be decided by experts, but not open to opinions and genuine choice."[36]

By making this decision, the treasure of freedom, democracy and *amor mundi* is lost.

THE POLIS AND *AMOR MUNDI*

At the end of *On Revolution*, after tracking the failure of the French and American revolutions to preserve freedom and politics, Arendt quotes Sophocles' "famous and frightening lines: 'Not to be born prevails over all meaning uttered in words; by far the second-best for life, once it has appeared, is to go as swiftly as possible whence it came.'"[37] Nietzsche would say that this line epitomizes nihilism. How can humans overcome this nihilism? "There [Sophocles] also let us know, through the mouth of Thesius, the legendary founder of Athens and hence her spokesman, what it was that enabled ordinary men, young and old, to bear life's burden: it was the *polis*, the space of men's free deeds and living words, which could endow life with splendor."[38]

Arendt's exaltation of the polis is her answer to Nietzsche. She agrees with Nietzsche that love is the answer to modern nihilism, but not Nietzsche's *amor fati*, love of (individual) fate, rather *amor mundi*, love of a world. How does *amor mundi* overcome nihilism? First, it rejects the atomism of modern individualism and instead connects us with each other, the world we live in, the ancestors who went before us and contributed to this world, and the children who will live in it later. Second, a world transcends us and gives us meaning. Meaning and significance is not "in our heads"—it is not something we make up or can alter at will according to our preferences. To be born into a stable world is to inherit certain ways of making sense of the world. To be clear this is not a monolithic set of beliefs. A world can have and probably always has multiple competing narratives. And while each individual has the option to take up and continue, to reject or to modify these meanings, they are meanings that are *in the world.* This represents a rejection of the idea that meaning and ethics are subjective, that is, that they are not real because they are just mental creations of humans. As we have seen, modern positivism casts doubt on anything that cannot be empirically verified, and as such modern philosophy was highly suspicious of ethics for a significant portion of the twentieth century. When existentialists rejected this scientism and the corresponding dismissal of questions of meaning and purpose, they nonetheless often did so in a way that Arendt thinks stressed individual meaning and individual responsibility.[39] The phenomenon of world preserves the reality of meaning and mattering while rejecting the tendency to make it a matter of individual choice and responsibility.

Third, *amor mundi*, especially when organized like the Greek polis, allows for the kinds of mutual recognition that serves as "public happiness." Arendt does not use the term "recognition" that has been popularized by both Charles Taylor and Axel Honneth.[40] The basic idea refers to Hegel's famous master/slave dialectic in *The Phenomenology of Spirit.*[41] In this allegory, two individuals fight for supremacy and to be recognized by the other. The winner of the battle, who becomes the master, actually fails to achieve recognition because by making the other a slave they consider them lower and incapable of recognizing them. Without delving into the dynamics of this dialectic, many contemporary thinkers have claimed that Hegel understood something important about humans: we want to be validated and recognized by others. Taylor and Honneth seek to use this insight about humans to think about the ways that exploitation is not just a matter of labor but also of social recognition. Those who are outcast, poor or given inferior status are denied the kind of recognition that humans seek. Writing before Honneth, Arendt does not use this Hegelian terminology, but it seems clear to me that the agonistic polis is about achieving honor, glory, and excellence—which not only can get the deeds and words of a particular individual recognized but also makes a particular world shine. The public happiness one feels in the polis is the happiness of knowing others see you, recognize your excellence and appreciate your contribution to the world. Public happiness is joy with others and joy with the world. When it works, the polis produces happiness and love.

As Chantal Mouffe argues, Arendt's portrayal of the agonistic polis has inspired many political thinkers. But Arendt herself argues that the agonism of the polis eventually tore it apart.[42] I have suggested that Arendt thinks there is an alternative to the agonism of the polis, and this is Socrates' community of friends. The Socratic version of the polis preserves the emphasis on public happiness, mutual recognition, and respect for the world, but rejects the agonistic, competitive nature of the polis in Athens.

This alternative is more important to Arendt that it might at first appear. We get a hint in a quote that occurs at the end of *On Revolution*. While discussing the importance of the polis she discusses the comments of a French resistance fighter from the Second World War, René Char:

> His book of aphorisms was written during the last year of the war in a frankly apprehensive anticipation of liberation; for he knew that as far as they were concerned there would be not only the welcome liberation from German occupation but liberation from the 'burden' of public business as well. Back they would have to go to the *épisseur triste* of their private lives and pursuits, to the "sterile depression" of the pre-war years, when it was as though a curse hung over everything they did: "if I survive, I know that I shall have to break with the aroma of these essential years, silently reject (not repress) my treasure." The treasure, he thought, was that he had "*found* himself," that he no longer suspected himself of "insincerity," that he needed no mask and no make-believe to appear, that wherever he went he appeared as he was to others and to himself, that he could afford "to go naked." These reflections are significant enough as they testify to the involuntary self-discourse, to the joys of appearing in word and deed without equivocation and without self-reflection that are inherent in action.[43]

To understand the full import of this requires knowing something about Arendt's personal history and personal struggles.

ARENDT'S HOPE: BULLSHITTING, STRATEGY, AND INTEGRITY

Arendt was a Jewish woman living in an anti-Semitic society. She was highly sensitive to the ways that the Jewish outsider often had to compromise their integrity to fit in. Arendt adopts the language of Bernard Lazare, who claims the Jewish outsider had two options: to be the Pariah or the Parvenu.[44] The Pariah can hold onto their traditions, but remains an outcast for so doing. The Parvenu, on the other hand, has to cast aside their traditions and seek to assimilate, to be like everyone else. But this is not a simple task, since often the outsider is recognized to be an outsider and they must double their efforts to cover over and erase this in order to fit in. Even if they can manage to pass themselves off as "like everyone else," this outsider status will often remain with them like an odor that can never fully be eliminated. They may always be "the Jew that doesn't act like a Jew and acts like us."

Arendt's first book was a biography of the German Jewish socialite and salon owner Rahel Varnhagen who lived from 1771 to 1833. Despite her outsider Jewish status, Varnhagen managed to ingratiate her way into high society. But in order to do so she had "to master the 'art of representing her own life: the point was not to tell the truth, but to display herself; not always to say the same thing to everyone, but to each what was appropriate for him.'"[45] In other words, she had to sell herself, to constantly seek to manage impressions, to play the game, to sacrifice her integrity, and become who they wanted her to be.

Arendt scholars have seen Arendt's interest in Rahel Varnhagen as a reflection of the tensions she dealt with in her own life as a Jewish woman.[46] Seyla Benhabib argues that Arendt's interest in Vernhagen reflects her interest in "a recovery of the 'public world' through authentic political action."[47] I think this claim is correct—that even in this early work, Arendt is thinking about world. Specifically, I think that Arendt is thinking about the ways that the modern world demands we assimilate into the giant economic family that she later would call "the social." Recall that she characterizes modernity as having "degenerated into mere administration, the public realm has vanished; there is no space either for seeing or being seen in action . . . political matters are . . . dictated by necessity to be decided by experts, but not open to opinions and genuine choice."[48] Instead of being free to act without guile, without masks, "to go naked" as Char claimed he could do during the rare moment of freedom that opened up during the revolutionary moment of resistance during the Second World War, modern society demands we jump through the right hoops, dance the right way, play the right games, and look the right part.

In others words, this is not just a problem for Jewish outsiders (although it is a more difficult problem for an outsider). Everyone in modern society must fit into the machine. We must play the game. As students we called this "BS-ing"—that is, doing the bullshit that one has to do to succeed, even if one thinks it is bullshit. Henry Frankfurt has written a famous monograph *On Bullshit* in which he basically claims that bullshitting is pretending to know what one doesn't know.[49] While it is true that BS-ing involves looking like one knows what one is doing even though one may not (especially BS-ing one's way through a test), it is more than this. Any time one does things that one thinks are bullshit in order to succeed, he or she is BS-ing. Often, it is used to refer to completing requirements or tasks that one thinks should not be required, but doing it anyway because it is required.

While everyone will likely have to do some BS, there are people who seem particularly willing to put up with this in order to get what they want, and this type of person is often referred to as a "BS-er." BS-ers tend to be successful. They ask the question: what do I need to do to get ahead? And they do it. Arendt was particularly interested in BS-ers. Varnhagen was a BS-er. Eichmann was a BS-er. This fascination on Arendt's part with BS-ers like Varnhagen and Eichmann was not because Arendt was a BS-er herself—although perhaps we all are in modern society to some degree—rather because Arendt is interested in the opposite of BS-ing, integrity.

Socrates was a model of integrity for Arendt, as he has been for many in Western thought. When she reconstructs Socrates' alternative polis, the community of friends, Arendt is attracted to and thinking about the possibility of a place that is free from BS and BS-ing. In her comment on the freedom of Char during the resistance, she speaks of "the joys of appearing in word and deed without equivocation and without self-reflection that are inherent in action."[50] Char had experienced a brief moment in his life in which he did not have to BS, in which he did not have to constantly think about how what he was doing looked to his bosses and his peers, about promotions, about impressions, and so on. It was in this sense that Char was able to unconscientiously "go naked"—to be himself, to say what he really thought, to "let it all hang out," to be honest, to be true to himself and with others.

This is freedom. Recall from *The Human Condition* that freedom means to be free from necessity. The private sphere, the sphere of the family, was focused on practical tasks of assuring the biological functioning of the home, making sure there is enough food, that children are cared for, and so on. In this setting there is little to no freedom, as the incessant needs of the family take priority. To be free means to be free from having to worry about issues of necessity—whether the boss who manages others or the workers who do what the boss says. The polis was supposed to be a second life, a space of freedom from necessity in which citizens could meet with each other. Arendt claims that the logic of the private sphere—the concerns of labor and work—had to be kept free from the polis or it would cease to be a space of freedom.

What is interesting about this passage that ends *On Revolution* is that Arendt is clearly stressing that this freedom is a freedom from "insincerity," from "masks" and from "make believe." She is pointing out freedom from the demands of a highly administered, image-obsessed modernity. This is the kind of world described in practical terms by the sociologist Erving Goffman. For Goffman, we are all actors, managing the image we give off, constantly on alert for how we are being perceived, always managing impressions. Strategy is a constant necessity in the dramaturgic social world of Goffman. Poor performance or strategic mistakes can lead to embarrassment, social reprobation and failure to move up the ladder. Great performance and careful strategic manipulation can lead to good social standing, better potential roles, and upward social movement.

This issue of image management and social strategy is one that does not occur in *The Human Condition*. Frankly, I wonder if a hypercompetitive polis would really be free from this kind of strategy, or—more likely—if the battle for fame and glory would necessitate it. In the passage on Char, Arendt claims that his discussion of being free from masks and impression management testifies "to the involuntary self-discourse, to the joys of appearing in word and deed without equivocation and without self-reflection that are inherent in action."[51] Here, she seems to be claiming that action involves freedom from these kinds of strategic manipulation. In other words, action (in Arendt's technical sense) is not just freedom from biological necessity, but freedom from the necessity of BS-ing and strategic manipulation.

In his *Theory of Communicative Rationality*, Habermas distinguishes between strategic action and communicative action.[52] The former refers to actions taken with the intent to manipulate a situation to get the results that one wants. Often, the desired result has been decided on prior to any interaction, and when such interaction occurs, the strategic actor attempts to steer the situation in the direction of their predetermined preference. Communicative action, on the other hand, approaches a situation without masks, without pretension, without a manipulative mentality. It is open minded, sincere and "naked" in the sense Char describes.

Whatever difficulties and shortcomings the *Theory of Communicative Rationality* may have, this fundamental distinction between strategic and communicative action seems to fit well with Arendt's concerns here. The superiority of Socrates' community of friends is not just that it creates a space for freedom and *amor mundi*, but that it avoids the necessity of strategic action that may be required in the agonistic polis. Since Socrates's community of friends is not agonistic but cooperative, it avoids the tensions that led to the breakdown of the agonistic polis.

Arendt's hope is for a space of freedom that is not just free from the demands of labor, but also the masks and make-believe of our modern highly administered society. Having personally experienced the demands of casting aside one's integrity to fit in, both as a Jewish outsider but also as human in modern consumer capitalism, Arendt seeks a rare space where one can "go naked" without masks and without bullshitting.

CONCLUSIONS: *AMOR MUNDI* FOR ARENDT

The polis of the Greeks is a concept of great importance for Arendt. It represents a kind of love of a world that we moderns have largely lost. The ancient polis was not perfect however and ultimately failed because it was torn apart by the agonism it promoted. Nonetheless, it does show us what it would be like to be committed to, honor and love a particular world.

While some scholars have been impressed by Arendt's invocation of the ancient polis and have been inspired to rethink modern democracy in light of it, I suggest that Arendt herself was more attracted to the Socratic alternative to the polis, his community of friends. This model retains the commitment to and love of the world of the polis without the antagonism that tore it apart. For Arendt, Socrates' community of friends is a kind of lost alternative, a lost treasure in Western thought. The West instead first followed late Plato's rejection of democracy in favor of *vita contemplativa* and the authority of the cosmos. When the Timaean cosmos fell apart, the Western tradition did not return to the politics of the polis or Socrates' community of friends, but instead followed the quantitative approach of early Plato—taking science as the primary modern authority. According to Arendt, this new *vita activa* of modernity has led to a pervasive modern world alienation. Now, living in a period of intense consumerism that has inverted the traditional Greek valuation of action over work and labor, we struggle to even recognize the phenomena of a stable world and instead move between fragile and multiple "worlds," ever-consuming and ever-accelerating.

Despite this world alienation, there have been historical moments when the lost treasure of freedom and politics has re-emerged briefly—especially during times of revolution in which questions of how to create a stable world have become live. The American Revolution in particular had a chance to recover space for politics and freedom. Ultimately, however, Americans chose economics over politics.

What are we supposed to do with this *amor mundi*? Because of the way Arendt places this lost treasure so far away from us, back at the beginnings of Western thought, it may seem easy to dismiss it as fanciful or impractical. But in her account, modern world alienation is a relatively new phenomenon in world history. While she does not examine other traditions besides ancient Greece, I would suggest there are other premodern societies that manifest some sort of *amor mundi*. In other words, there have probably been many city-states, communities, and peoples that have had an intense love of world. Indeed, with the help of historians and anthropologists, it would be possible to compare and contrast various forms of premodern *amor mundi*—I suspect there are many that should be of interest to us.

I want to examine one such alternative: *amor mundi* as it was experienced by Native Americans. In the next chapter, I turn to Vine Deloria Jr. and Daniel Wildcat to help think through a different kind of *amor mundi*—love of the Earth/Nature/place.

NOTES

1. http://pavilionmagazine.org/chantal-mouffe-agonistic-democracy-and-radical-politics/. Accessed January 1, 2019.
2. Ibid.
3. Arendt, Hannah. "Philosophy and Politics." *Social Research*, Vol. 57, No. 1 (Spring 1990), 73.
4. Arendt, "Philosophy and Politics."
5. Ibid., 74.
6. Ibid., 78.
7. Ibid., 74.
8. Ibid., 80.
9. Ibid., 80.
10. Ibid., 81.
11. Ibid., 81.
12. Ibid., 82.
13. Ibid., 83.
14. Ibid., 84.
15. Ibid., 82.
16. See Chapter 2 of Arendt, Hannah. *The Human Condition* (Chicago: University of Chicago Press, 1998).
17. Arendt, *The Human Condition*, 28–29.
18. Ibid., 29.
19. Ibid., 38.
20. See Tucker, ed., *The Marx and Engels Reader (Second Edition)* (New York: W.W. Norton & Company, 1978); Castro-Gómez, Santiago. "The Social Sciences, Epistemic Violence and the Problem of the 'Invention of the Other.'" *Views from South*, Vol. 3, No. 2, 2002, p. 269–285.; Foucault, Michel. *Discipline and Punish* (New York: NY: Vintage Books, 1995) and Foucault, Michel. *Power/Knowledge* (New York: Vintage, 1980).
21. Arendt, Hannah. *On Revolution* (New York: Penguin Classics, 2006), 62.
22. Ibid., 64.
23. Ibid., 63.

24. Arendt, *The Human Condition*, 4–5.
25. Ibid., 125–126.
26. For more on this see Arendt, Hannah. *The Life of the Mind* (New York, NY: Harcourt Brace Jovanovich, 1978). I have discussed this issue in detail in my book Pack, Justin. *How the Neoliberalization of Academia Leads to Thoughtlessness: Arendt and the Modern University* (New York: Lexington Books, 2018).
27. Arendt, *The Human Condition*, 5.
28. Ibid., 5.
29. Arendt, *On Revolution*, 21–22.
30. Chapter 2 of *On Revolution* discusses the tensions of this project.
31. Ibid., 35.
32. Ibid., 126.
33. Ibid., 128.
34. Ibid., 141.
35. Ibid., 133.
36. Ibid., 237.
37. Ibid., 281.
38. Ibid.
39. Arendt, Hannah. "Culture and Politics." *Reflections on Literature and Culture* (Stanford, CA: Stanford University Press, 2007).
40. Honneth, Axel. *The Struggle for Recognition: The Moral Grammar of Social Conflicts* (Boston: The MIT Press, 1996).
41. Hegel, G. W. F. *The Phenomenology of Spirit* (New York: Oxford University Press, 1977).
42. Arendt, "Philosophy and Politics."
43. Arendt, *On Revolution*, 280.
44. Pitkin, Hannah Fenichel. *The Attack of the Blob: Hannah Arendt's Concept of the Social* (Chicago: University of Chicago Press, 1998), 21.
45. Pitkin, *The Attack of the Blob*, 26.
46. Moruzzi, Norma. "From Parvenu to Pariah: Hannah Arendt's *Rahel Varnhagen*," in Fleischacker, Samuel, ed., *Heidegger's Jewish Followers* (Pittsburg, PA: Duquesne Press, 2008.
47. Benhabib, Seyla. "The Pariah and Her Shadow: Hannah Arendt's Biography of Rahel Varnhagen." *Political Theory*, Vol. 23, No. 1 (Feb. 1995), 7.
48. Arendt, *On Revolution*, 237.
49. Frankfurt, Henry G. *On Bullshit* (Princeton: Princeton University Press, 2005).
50. Arendt, *On Revolution*, 237.
51. Ibid.
52. Habermas, Jürgen. *The Theory of Communicative Rationality, Vol. 1, Reason and the Rationalization of Society* (Boston, MA: Beacon Press, 1984).

Chapter 5

Love of the Earth in Native American Philosophy

In chapters 1 and 2, I examined one kind of *amor mundi*: love of the cosmos. I claimed that love of the cosmos was predominant in Western thought after its widespread adoption from Plato into the cosmologies of the Judeo-Christian-Islamic traditions up until Copernicus and the rise of modernity.

In chapters 3 and 4, I examined a second kind of *amor mundi*: love of a (political) world. I examined Arendt's claim that this kind of love existed in the Greek polis and in Socrates' alternative to the agonistic polis. According to Arendt, the turn to the cosmos with Plato later deemphasized this kind of *amor mundi*, but it really disappeared with modern world alienation. She claims modern science and modern consumerism have greatly diminished the phenomenon of a stable world. Despite this, there have been moments where the hidden treasures of freedom and politics have re-emerged—especially during revolutionary moments. Arendt's purpose is to recover a sense of what is meant by world and illuminate the thoughtlessness of our contemporary world alienation.

While Arendt focuses on the ancient Greek polis as a key moment of *amor mundi*, a comparative historical and anthropological analysis would reveal that there are many city-states, communities, and peoples that demonstrate/d love of their particular (political) world.

In this chapter, I want to focus on a third form of *amor mundi*: love of the Earth/Nature/place. To do so, I turn to Native American philosophers Vine Deloria Jr. and Daniel Wildcat. The reader may have noticed that I have employed the awkward triad Earth/Nature/place to prevent the reduction of Native American *amor mundi* to love of the Earth or love of Nature, which, as we will see, Deloria Jr. claims tend to be misunderstood in Western interpretations of Native American thought. More on this in a moment. I believe that of the three kinds of *amor mundi* I have been examining, love of the

Earth/Nature/place is the most accessible. Love of the cosmos is difficult to recover because the Platonic metaphysics that it was traditionally built on has been discredited. Love of a particular political world is difficult for a variety of reasons, but could be recovered. Love of the Earth/Nature/place is perhaps the easiest of the three to try and make a living reality—in fact some people are already doing it.

A FEW CAVEATS

Before turning to my analysis, let me point out two difficulties with the topic of this chapter. The first concerns generalizations about Native Americans—in this case made by Native Americans. Obviously, there are many different groups that are "Native American," and they may have radically different histories, customs, traditions, ethics, and so on. There is a danger in generalizing and speaking about "Native Americans" as if they are all roughly similar. And yet Vine Deloria Jr. will consistently speak about "Native Americans" or "American Indians" in contrast to Western thinkers, Western science, and so on (we could also worry about generalizations about "Western science" or "Western thought"). This is because Deloria, while academically trained, is also an activist concerned with promoting much-needed social change. He knows that there are different groups of Native Americans and that there are different strands of Western thought—including some which support his own position,[1] but the speed and scope of destruction of Native people and the Earth requires a radical reorientation now. Generalizing about "Native Americans" and "the West" allows for a stark contrast that may be necessary to *see* the causes of this destruction and the depth of these problems. Deloria would recognize that particular cases require an analysis that is related to a particular place—indeed, this is required by his metaphysics—but in terms of understanding the general problem, he paints in broad strokes.

The second caveat concerns the appropriation of Native American thought by non-Native Americans. There is a long history of European thinkers marshalling Native Americans for their own narratives—whether to portray Native Americans in a negative light in order to claim some sort of superior position or in an idealized light in order to critique European traditions (a classic example of this is Rousseau). Demonizing or romanticizing Native Americans are both manipulative uses of Native Americans for a purpose determined by and for someone else.[2] Furthermore, even nonindigenous scholars who seek to listen to Native American thinkers and popularize Native American perspectives, risk profiting off them—in other words, for example, a white advocate may advance their career by discussing Native

American traditions, thought, or art, while Native Americans themselves may find they cannot advance in certain careers except by acting white or being supported by some kind of white patron. In *Power and Place*, Deloria comments that "without the voices of respected white scholars, there is little chance that we can get sufficient attention from the scientific establishment in order to plead our own case."[3]

I am a white scholar attempting to sympathetically present Deloria and Wildcat in this chapter. My claim is that what Deloria and Wildcat present as Native American metaphysics functions as a third kind of *amor mundi*: love of the Earth/Nature/place. I hope that my presentation of their arguments would be something they would approve and fall under the category of attempting to help get sufficient attention to their claims, instead of using Native American thought for my own purposes. I would hope readers that find the issues presented in this chapter compelling will go and seek out the works of Deloria and Wildcat and other Native American thinkers (but, Deloria insists, they should also think hard about their own culture and the damage it is doing instead of looking at Native American cultures as a potential savior).

I have employed the awkward triad Earth/Nature/place precisely to try and avoid one such misunderstanding. In a piece entitled "Kinship with the World," Deloria says the following:

> For the most part Indians do not "deal with" or "love" nature. In the Western European context human experience is separated from the environment. When Indians are told that they "love nature," they cannot deal with this because nature is not an abstraction to them.
>
> Indians do not talk about nature as some kind of concept or something "out there." They talk about the immediate environment in which they live. They do not embrace all trees or love all rivers and mountains. What is important is the relationship you have with a particular tree or a particular mountain.[4]

According to Deloria, one of the key differences between Native American and Western thought is that the former emphasizes experience, while the latter emphasizes concepts.[5] Already, then, there is a danger in using the concept "*amor mundi*" to describe Native American relationships with the Earth. Deloria points out that white readers of Native American thinkers often come seeking for a non-Western and more ethical way of relating to the Earth, but they tend to read Native American experience through Western conceptual filters. "When we talk with non-Indians about nature, there is really nothing that we can say in universal Western concepts that is going to make a lot of sense."[6] This creates a strange situation for young Native Americans who are being educated in Western schools:

> In the modern period, Indians are in tremendous transition because we are going through your educational system, wherein we move away from the specifics of the universe traditional Indians lived with and we move toward the scientific way of thinking. Many Indian people are leaving their culture and traditions in ways they do not suspect. They are developing a schizophrenia. They look at their grandfather, who goes out and talks with birds and coyotes, and they think he is superstitious. And yet they go to school and they learn that they are supposed to love nature and learn from nature. So we are getting, in my opinion, a generation of lost Indians who are going out there, just like lost whites, and trying to embrace trees and think this is doing something Indian. Lost whites come to the West to love the environment, and they end up paving the damn thing and subdividing it. . . . Certainly, I would like to keep all of them east of the Mississippi, and let them clean up what they have already screwed up.[7]

Deloria rejects the idea that there is some broad, overarching love of Earth in Native American traditions. The perception that such a thing exists is a result of lost whites misinterpreting the specific love of particular Native Americans for specific plants and animals as a kind of general love of the Earth or nature. For Deloria, the problem for lost whites is their own culture: "You have got to look back into your own culture . . . when you start working your way through that, I think you will see more of what we are talking about."[8]

I don't think Deloria is saying that lost whites and Native Americans cannot communicate, that they have incommensurable paradigms or that whites should never seek to learn from Native Americans. He declares frankly in *Power and Place* that Native American metaphysics is superior to (his reading of) Western metaphysics. Clearly, he wants unhealthy Western practices to stop and thus, his activism and writing to both American Indians and non-Indians. Furthermore, he claims it often takes sympathetic white allies in positions of power to get the Western establishment to consider listening to Native American perspectives.[9] But given the historical tendency in Western thought to place itself as the culmination of human progress, Deloria's exhortation to "clean up what they have already screwed up" is well taken.

There is no foolproof method for assuring that I don't make some of these mistakes in my discussion of Deloria and Wildcat. The highly conceptual work of Arendt and Native American emphasis on experience may simply be difficult to "fit." With that said, Deloria is himself making the effort to describe Native American thought in conceptual terms in *Power and Place*. My hope is that I can avoid the worst of the lost white approach in presenting Deloria and Wildcat.

ARENDT AND EARTH: IS ARENDT ANTHROPOCENTRIC?

The impetus that led to this book was reading Arendt's *The Human Condition* alongside Deloria and Wildcat's *Power and Place: Indian Education in America.* As I discussed in chapter 3, Arendt makes a distinction between world and Earth. Interestingly, she begins *The Human Condition* with a discussion of Earth alienation—specifically, with the desire to escape the Earth through space travel she saw expressed at the time of the launch of Sputnik. Arendt then points out that despite Mother Earth being a condition for the possibility of human existence, there is an odd resentment directed at her. When read alongside Deloria and Wildcat, it occurred to me that her concerns are very similar to some of the claims they make and I could imagine them being pleased with her discussion. But after using the Earth to illustrate this strange resentment of the condition of human existence, Arendt then turns primarily to the phenomenon of (human) world. After this initial discussion of Earth, the issue of Earth alienation largely drops from the text and the focus becomes world alienation. I could imagine Deloria and Wildcat being disappointed that Arendt did not follow through with the thread of Earth alienation.

What does Arendt miss by focusing on world instead of Earth? In some ways, her critique of modern world alienation parallels the critique offered by Deloria and Wildcat. All three agree that modern world and Earth alienation share the idea that a fundamental condition of human existence has been covered over and as a result is often ignored. Clearly, Arendt is aware that not just world, but Earth, has been forgotten. Nonetheless, her focus remains on world. Does this perhaps reflect some sort of fundamental anthropocentrism on her part? In other words, does her attention to world over Earth reflect a privileging of the human over the nonhuman?

It very well may, but it is important to note that Arendt's phenomenology of world is itself a move that can be construed as attempting to overcome anthropocentrism. World, for Arendt, transcends humans. A stable world (if we are born into one) existed before us and will likely exist after us.[10] It makes demands on us and shapes our existence in ways that we do not control and may not even realize. When Arendt critiques world alienation and lauds the Greek polis or Socrates' community of friends, she is pointing away from modern individualism toward something transcendent: world. Thus, from the perspective of modern individualism, Arendt is pointing to a phenomena that is not anthropocentric. World, in some important ways, is not human.

World, however, occurs in and on a greater whole, Earth. From Deloria and Wildcat's perspective, it must be disappointing that Arendt did not follow up on her insights about Earth alienation. Furthermore, from the perspective of Earth, a particular human world is centered on humans. Native American metaphysics includes animals and plants in the community that makes up a particular world, but Arendt never discusses animals and plants. Thus, while her discussion of world may seek to overcome modern individualism and seems to push back against anthropocentrism on one level, from a Native American perspective it remains thoroughly anthropocentric.

This does not mean, however, that Arendt's genealogy of world alienation would not be of interest to Native American thinkers. The rise of modern world alienation is very similar to modern Earth alienation: both forget something that is fundamental to the human condition.

I want to look at how Native American metaphysics moves beyond a particular human world to the Earth and provides a rich account of place.

NATIVE AMERICAN METAPHYSICS VERSUS WESTERN MACHINE METAPHYSICS

According to Deloria and Wildcat, there are critical differences between Native American Metaphysics and Western metaphysics. Like we saw in chapter 2, this is largely a result of modern science. According to Wildcat, science tends to reduce reality to a physical world that can be understood by breaking into the smallest pieces similar pieces, whether these are atoms or genes, and so on. This attempt to dissect the world around us in these smallest pieces is worryingly reductionist and problematically shapes our understanding, our methods, and our understanding of reality.[11] Like Brague, Jonas, Nietzsche, and Merchant, Wildcat claims this approach kills the living cosmos and replaces it with a dead, mechanical one. This results in a kind of "popular mechanics" view of the world that seeks to find these basic pieces of reality, understand how they fit together and potentially how they could be rearranged. Wildcat doesn't claim that this approach doesn't work, just that it only allows for understanding one part of reality:

> Let's be clear: certain "things" can be understood using the metaphysics of time, space and energy. However, a great deal of what we experience cannot be explained within the metaphysics of Western science, and that is the critical point. An entire realm of human experience in the world is marginalized, declared unknowable, and, consequently, left out of serious consideration. This reality cannot fit in the objective experimental box of mechanical cause and effect, and no tool or technology will change this situation unless we merely say that all there really is to the world is mechanics (in the structural sense) and tools. Western notions of reality and corresponding ideas of knowledge are not far from this cold "scientific" assessment.[12]

Wildcat recognizes that science is indeed a powerful way to understand the world around us. But it would be a mistake to think that a scientific account of reality describes all of reality. Both Deloria and Wildcat stress over and over that science and Western metaphysics are reductive: they only describe one aspect of reality. The basic counterclaim is straightforward: human experience is far richer than the interplay of forces, atoms, and energies. Everyday experience, on the contrary, is full of meaning and significance. The humans I interact with are not bones, muscles, and skin but my family, friends, coworkers, and so on. Thus, while science may indeed describe the physiological aspect of the world around us, it would be an odd and terrible mistake to take this as the "true" or only reality. To do so risks skipping over the most important aspects of human experience, and the result is a "true" reality that can do very little to help us understand human experience. Wildcat correctly points out that science has to declare important aspects of human experience as "unknowable" or subjective. This is particularly important in light of Native American metaphysics because for Deloria and Wildcat, the aspects of reality that are most important to Native American experience are excluded from the realm of the knowable and therefore the real in the Western approach. We can understand this better by turning to Native American metaphysics.

According to Deloria the concepts that are needed to understand Native American metaphysics are power and place:

> Power being the living energy that inhabits and/or composes the universe, and place being the relationship of things to each other. It is much easier, in discussing Indian principles, to put these basic ideas into a simple equation: Power and place produce personality. This equation simply means that the universe is alive, but it also contains within it the very important suggestion that the universe is personal and, therefore, must be approached in a personal manner.[13]

The metaphysics here is the opposite of a dead, mechanical universe—on the contrary, everything is alive. This kind of claim may not be completely foreign to someone coming from a Western background, as there are strands of Western thought that also claim the cosmos is alive (as we have seen). Deloria stresses that this extends beyond animals and plants, however—power is in rocks, rivers, landscapes, wind, weather, and so on. Furthermore, Deloria claims that many Native American creation stories portray humans as the last creation and therefore the younger siblings of the other beings that were created before us.[14] This implies that plants, animals, rocks, and landscapes are the older brothers and sisters of humans and that we can and should learn from them.

The implications of Native American metaphysics are far-reaching and explain why Deloria and Wildcat are so emphatic in their denunciation of reductive Western metaphysics. First, if the universe is alive and the objects around us in the world are the older brothers and sisters of humans, then the human quest for understanding and wisdom must turn to them—after all, they have more experience than we do. As such, Deloria claims that Native American epistemology is directed at studying and learning the natural world. But this is not the scientific study of what the natural world is composed of, but how the living world acts and interacts. Older brothers and sisters act in certain ways because they have experience and know what works in certain situations. By observing these patterns, Native Americans can gain knowledge about what one should do in specific places in interaction with the other interconnected living beings that exist together in that place.[15] Native American technologies are based on working with the plants, animals, soils, weather, and features that are relevant to a particular location.[16] This involves being respectful and learning from life: "A great gulf exists between these two ways of handing knowledge. Science forces secrets from nature by experimentation, and the results of the experiments are thought to be knowledge. The traditional people accepted secrets from the rest of creation."[17]

This brings us to a second feature: if the universe is alive, and we can learn from it, then epistemology is intimately tied to ethics. The older brothers and sisters are willing to share their insights and experience, but our relation with them must be, as Deloria says, personal. Obviously, this is not just a point about epistemology, but a broader point about relationships between humans and the world around us: we should respect and listen to the world around us. From a Native American perspective, one of the most disturbing aspects of Western metaphysics and the ensuing practical interaction with nature is the utter lack of recognition of life of the natural world and corresponding disinterest in listening to or respecting nonhuman peoples. The dead universe is available for use by and for humans. While both modern science and American Indian traditions ask "(1) 'How does it work?' and (2) 'What use is it?'" American Indian traditions also ask "What does it mean?"—which for Deloria is an ethical question that implies further questions about appropriate ethical relations with it.[18]

Particularly important to this epistemology and ethics is the third feature: the role of place. Here we can begin to truly understand how radically different Western metaphysics and science is from Native American metaphysics, epistemology, and ethics. Knowledge and ethics on this model are tied to place. This is because knowledge depends on the particular peoples and their interrelationship with each other in a specific place. The same holds for ethics. Since knowledge and ethics are tied to place, this suggests that different places have different kinds of knowledge and different kinds of ethics. The Sonora desert in Arizona has a unique set of peoples (animals, plants, soils, landscapes, etc.) that make up that place and they relate to each other in a way that is unique to that place. As a human, my knowledge and behavior should respond to these peoples with regard to both their particular unicities and their relationships with each other. The Coastal Redwood forests of Northern California have different peoples and therefore my knowledge and behavior would be different there. On this model, ethics is not a set of generalizable principles that are universalizable to all human interactions. On the contrary, ethics is specifically tied to place. The same would follow for knowledge. The Rufus hummingbird will act differently in the Sierras than it might on the Mogollon rim. It cannot be reduced to a set of abstract features (bone structure, color, reproductive habits) but must be understood in connection with place.

PLACE, SIBLINGS, AND SPIRITUALITY

It is worth taking a moment to attempt to appreciate what it would be like to experience the world this way. Like the premodern Western experience of the Timaean cosmos and the premodern experience of a particular human world, the kind of relationship to place and the Earth that is being described here involves a kind of "porousness" and connection that is radically different from modern experience. Take, for example, the opening passage from the book *Braiding Sweetgrass: Indigenous Wisdom, Scientific Knowledge and the Teaching of Plants* by the Native American biologist Robin Wall Kimmerer:

> Hold out your hands and let me lay upon them a sheaf of freshly picked sweetgrass, loose and flowing, like newly washed hair. Golden green and glossy above, the stems and banded with purple and white where they meet the ground. Hold the bundle up to your nose, find the fragrance of honeyed vanilla over the scent of river water and black Earth and you understand its scientific name: *Hierochlie odorata*, meaning the fragrant, holy grass. In our language it is called *wiingaashk*, the sweet-smelling hair of Mother Earth. Breathe it in and you start to remember things you didn't know you'd forgotten.[19]

This beautiful passage invokes a relationship to sweetgrass that is hard for me personally to access. Before reading it, I probably would have seen sweetgrass and thought little more than "that is a weird grass" or "that smells nice." For Robin Wall Kimmerer, there is a depth of connection with this grass that I find difficult to imagine—and yet this passage seeks to help us experience it also—to see grass as something special and holy: "Will you hold the bundle while I braid? Hands joined by grass, can we bend our heads together and make a braid to honor the Earth? And then I'll hold it for you, while you braid too."[20]

The passage here is infused with love. Wall Kimmerer invites us to relate to grass, and through grass to each other. We get a sense everything being connected. As Wildcat puts it: "We, human beings, in all our rich diversity, are intimately connected and related to, in fact dependent on, the other living beings, land, air, and water of the Earth's biosphere. Our continued existence as a part of the biology of the planet is inextricably bound up with the existence and welfare of the other living beings and place of the Earth: beings and place, understood as persons possessing power, not objects."[21]

Robin Wall Kimmerer seeks to help us *see* the grass and through the grass our connection to everything. Grass is such a small thing—so easily dismissed and ignored. I doubt most of us can even distinguish different kinds of grass, much less relate to them/it as a person. Another small being that she has drawn attention to is moss. In her *Gathering Moss: A Natural and Cultural History of Moss* she uses both science and her cultural traditions to get us to *see* something that we would usually ignore.[22]

Understanding and relating to grass and moss involves understanding soil, precipitation, seasons, other plant competitors, fungi, the animals that eat and interact with them, and so on. In this sense, understanding grass and moss means understanding all the many things that are connected to them in a particular place. Deloria resists concluding from this that Native Americans love the Earth, but he does so to resist this being understood as something abstract. Instead, this connection must be understood as a connection to the specific peoples of a particular place.

E. Richard Atleo of the Nuu-chah-nulth of western Vancouver Island stresses that this interconnection must also be understood as having a fundamental spiritual dimension. He proposes "an indigenous theory" based on the principle that "everything is one"—*heshook-ish taswalk* in Nuu-chah-nulth.[23] Based on Nuu-chah-nulth traditions, he claims that everything in the physical realm derives from a more fundamental spiritual one. Unlike Christianity, which claims a superiority for humans, Nuu-chah-nulth traditions claim that all life has a common origin and that, as such, no particular people is superior to any other.[24] Furthermore, these life forms commonly transform into other forms—humans into salmon, spiritual beings into humans, and so on.[25] While Western practices tend to stress individualism, Nuu-chah-nulth traditions stress community, both with other humans and with life forms in a place.[26]

Compared to Western morality, which is largely concerned with human relations with other humans, American Indian morality is much more crowded. Proper relationships occur not just between humans, but between humans and other peoples. How are the relations to be established and maintained? Atleo claims that this is done through traditional stories and religious ceremonies. Traditional stories teach about proper balance and harmony and show both good and bad examples of how to act. Appropriate ceremonies, prayer, and action establish and preserve proper community relations.

Deloria and Wildcat agree with Atleo that the connection between humans and other nonhuman peoples is spiritual. As the Osage theologian George Tinker says:

> Most adherents of traditional American Indian ways characteristically deny that their people ever engaged in any religion at all. Rather, these spokespeople insist, their whole culture and social structure was and is still infused with a spirituality that cannot be separated from the rest of the community's life at any point. Whereas outsiders may identify a single ritual as the "religion" of a particular people, the people themselves will likely see that ceremony as merely an extension of their day-to-day existence, all part of which are expressed within ceremonial parameters and shall be seen as "religious."[27]

Wildcat approves of the claim Tinker is making here that there is a wholeness to Native American experience that defies the reductive categories of Western thought. Atleo, Wildcat, and Deloria all claim that this wholeness is missed by Westerners, because Westerners tend to separate experience into specific categories, for example, spiritual and physical, eternal truth and transitory experience, science and politics. As a result, when Westerners approach Native American thought and attempt to understand or imitate it, they do so through a fragmented lens that divides the world up in ways that do not fit with Native American experience. This impedes accurately understanding Native American practices—when sifted through a Western lens, these practices are warped in ways that reflect Western intellectual categories and Western metaphysics.

The fundamental issue at stake here is not whether Native American traditions are accurately represented however. Rather, the issue is the destruction caused by Western practices. This involves not only the destruction of Native American peoples but also the systematic and steady destruction of the environment. Not surprisingly, some of the major themes of Native American writing concern Native American sovereignty and the preservation of Native American traditions in light of the systematic attempt to eliminate them.[28] Important works like Winona LaDuke's *Recovering the Sacred: The Power of Naming and Claiming* function on two levels: one aimed at inspiring Native Americans and another to speak to the dominant white culture that has caused this destruction. LaDuke relates histories of the persecution of Native Americans and accounts of their contemporary attempts to recover their tribal practices and revitalize traditional relations with a particular place. In one such account, she tells the story of Mt. Graham, a sacred site for the Apache, but also the site of telescopes constructed by the University of Arizona. LaDuke narrates how the University of Arizona constructed the Large Binocular Telescope despite the objections of the San Carlos Apaches. In another, she relates the fight of the Ojibwe in Minnesota to grow wild rice. This wild rice was a part of traditional Ojibwe food in Minnesota but has also proven profitable for various companies who can sell it as a "natural" or more organic rice. Wild rice was grown in California but marketed with images that tied it to Minnesota Native Americans, thus, profiting off these tribes with both their traditional rice and their image.[29]

These and other accounts are meant to show that the sacred can be recovered and rally Native American communities to these causes, but also offer a critique of the continued destruction caused by the dominant culture. This critique of Western thought and practice is simultaneously aimed at Native Americans and Westerners: Native Americans are encouraged to resist losing their traditions to the dominant culture and the dominant culture is encouraged to change its ways. This pattern of speaking to two audiences at the same time is seen in many Native American works, including *Power and Place*. Deloria and Wildcat are speaking not only to Native American students but also to Western readers.

The message for the dominant culture is clear: let Native Americans have sovereignty and make their own choices, and destructive Western practices must change. Again, this destruction concerns not just how Native Americans are treated but also concerns environmental destruction. For Atleo, Deloria, Wildcat, and Robin Wall Kimmerer, the current environmental crisis is a clear sign of "imbalance and disharmony" (Atleo) that, according to Deloria, results from the critical flaws of Western metaphysics.

"YOU HAVE GOT TO LOOK BACK INTO YOUR OWN CULTURE": SCHIZOPHRENIC WESTERN METAPHYSICS

It is understandable why ecologically concerned "lost whites" would be attracted to Native American thought. Native American metaphysics and traditional practices seem to respect the environment in a way that is largely absent from most Western practices. For Deloria, the environmental crisis cannot be solved by Native Americans alone, however—it will take a radical change in Western practices, ethics, epistemology, and, most fundamentally, metaphysics. Thus his claim that "you have got to look back into your own culture."[30]

For Deloria, the destructive practices of the West result from its flawed metaphysics. His basic claim is that Western metaphysics leaves out fundamental aspects of human experience and then treats the knowledge it gains about the remaining part as the only or best way to understand reality. In "Kinship with the World" he argues that key metaphysical divisions in Western understanding fragment reality in a way that leaves Western knowledge with only, at most, a quarter of human experience. The first division of reality is between this realm and another "truer" realm. He agrees with the claim made by Karl Jaspers and later reasserted by Robert Bellah and Charles Taylor that during what Jaspers called the "Axial Age" many "urban civilizations" posited the existence of some sort of transcendental realm—whether something like the realm of the forms, heaven, or some other kind of more fundamental reality.[31] This other realm was often articulated as a real or true world, and this world and our everyday experience was considered transitory, false, or unimportant. What really mattered was getting back to heaven or accessing this other realm somehow. For Deloria, this creates an odd denial of this world and devalues our everyday experience here in favor of some other reality. In addition to this, he claims that Greek philosophy further divided the world into mechanical nature on one side and the political world of human civilization on the other. Human civilization is pitted over and against nature and the non-Greek barbarians. This results in a "pessimistic attitude towards the world [i.e., everything including nature and other humans outside of Greek city states] and a desire to go out imperialistically conquer new lands and new people."[32] When the Greek division between human civilization and nature is combined with the Axial Age emphasis on another otherworldly realm we get "a view of the world that splits it twice: once in terms of eternal verities as opposed to transitional experiences; and once in which it divides the world between science and political affairs."[33]

With the rise of modern science, the emphasis shifted back toward this world and empirical reality, but in a way that nonetheless retained a division between the inner workings of the world as revealed by science and the apparent world in which we live our everyday life. As a result:

> By the nineteenth century, we had grabbed that quarter chunk of the world of science and said, "We are going to investigate all of human reality, or all of world reality, and we are going to being interpreting all experiences according to the workings of the physical world. We're going to create social science in which we can treat our own selves and our own psyches as if they were something objective that could be observed and described scientifically."
>
> We have reduced our knowledge of the world and the possibility of understanding and relating environment to a wholly mechanical process. We have become dependent, ultimately, on this one quarter of human experience, which is to reduce all human experience to a cause-and-effect situation.[34]

Whatever we make of potential flaws of Deloria's historical account, his basic claim is that the modern scientific understanding of reality cuts out some of the key parts of everyday experience. The power of science comes by radically limiting what science studies to the quantifiable aspects of the physical universe. As Wildcat points out, science is good at telling us how this part of reality works.[35] This is only a small part of human experience however—and yet positivism and/or scientism insist it is the only form of legitimate knowledge. Deloria thinks that this is a warped and reductive view of reality that does not square with human experience. This is not a claim that is unique to Deloria, indeed phenomenologists and existentialists have made this a centerpiece of their critique of Western philosophy and culture also. Thinkers like Nietzsche, Dostoyevsky, Heidegger, Husserl, Jonas, and so on, have claimed that the mechanical account of life is inaccurate to human experience. We experience life as full of meaning and existential import, not merely as the interplay of forces and mechanical movements. Unfortunately, Western thought often treats values and meaning as subjective preferences. In so doing, value and meaning are placed in the human mind and are not "out there" in the world. Phenomenologically however, this is incorrect, this not how we experience life—on the contrary, in human experience the world is full of meaning.[36]

When you combine the prestige and success of science with the "subjectivizing" of values and meaning in Western thought, you get a model that tends to dismiss values and meaning as unscientific and less real and are left with only material reality. As we saw in chapter 2, this not only results in a mechanical, "dead" universe but also leaves morality without a foothold in reality. We have already examined these ideas in thinkers like Nietzsche, Jonas, and Merchant.

Deloria is fully supportive of the critiques of Western thought and practice made by Western thinkers. In fact in *The Metaphysics of Modern Existence*, his strategy is to catalogue and present these critiques which he takes to indicate a kind of cumulative breakdown of the flawed model of modern science:

> In the fields of scientific knowledge and social reform we see a gradual and irreversible movement away from the sterility of the traditional Western European formulas and doctrines into a more flexible and broader awareness of the manifold experience of life . . . the view of the world which formerly dominated Western peoples and which currently dominates Western science is being transformed into an ancient and all-encompassing attitude towards life, best characterized by the American Indian cultures and traditions.
>
> Over a period of time it became increasingly clear that the trend in modern thought was approaching the Indian conception of the universe.[37]

This is a remarkable claim that arguably has not been borne out (not yet at least). Deloria wrote this in 1979, at the end of a decade that saw rising awareness of the environmental crisis and increased visibility of Native Americans as a result of the American Indian Movement. But it was also the same year that Margaret Thatcher was elected prime minister of England and the year before Ronald Reagan was elected President of the United States. This ushered in an era of neoliberal capitalism that has marshaled science into a source of new profitable technologies and made it even more difficult to criticize its flawed ontology.[38]

Deloria recognized that some readers would be frustrated to find a Native American writer focused on a synthesis of Western arguments instead of presenting his "authentic" Native American position more straightforwardly.[39] In the preface to the 2012 republication of *The Metaphysics of Modern Existence*, Wildcat claims that the book remains a very important one because Deloria seeks to offer an articulation of Native American metaphysics, even if it is by way of a synthesis of marginalized Western critiques. But for Deloria, these marginalized critiques are important to catalogue and synthesize not just because he claims they are pointing toward an understanding of the world that is "best characterized by the American Indian cultures and traditions," but also to show that there are many reasons to reject the hegemony of modern science. Note that, like Wildcat, Deloria is not saying science is wrong, but that the cultural tendency in the West to make science hegemonic and to treat it as the primary form and pinnacle of knowledge is deeply misguided: "Science and technology reign today as the practical gods of the modern age; they give us power to disrupt nature but little real insight into how it functions."[40]

One of Deloria's favorite critics of this kind of scientism is Paul Feyerabend. He claims Feyerabend's "work is critically important for non-Western and post-Western peoples" because he "is a threat to the routine operation of philosophy, science, and the troves of accumulating human knowledge because he asks penetrating and embarrassing questions in fields that most people feel have been laid to rest."[41] In other words, Deloria appreciates Feyerabend's ability to disrupt and challenge the hegemony of Western thought.

It is easy to see why Deloria would like Feyerabend, as the latter repeatedly condemned the idolization of science:

> Scientific "facts" are taught at a very early age and in the very same manner in which religious "facts" were taught only a century ago. There is no attempt to waken the critical abilities of the pupil so that he may be able to see things in perspective. At the universities the situation is even worse, for indoctrination is here carried out in a much more systematic manner. Criticism is not entirely absent. Society, for example, and its institutions, are criticised most severely and

> often most unfairly. . . . But science is excepted from the criticism. In society at large the judgment of the scientist is received with the same reverence as the judgement of bishops and cardinals was accepted not too long ago. The move towards "demythologization," for example, is largely motivated by the wish to avoid any clash between Christianity and scientific ideas. If such a clash occurs, then science is certainly right and Christianity wrong. Pursue this investigation further and you will see that science has now become as oppressive as the ideologies it had once to fight. Do not be misled by the fact that today hardly anyone gets killed for joining a scientific heresy. This has nothing to do with science. It has something to do with the general quality of our civilization. Heretics in science are still made to suffer from the most severe sanctions this relatively tolerant civilization has to offer.[42]

Feyerabend also shares Deloria's frustration with abstractions:

> I started examining the rise of intellectualism in Greece and the causes that brought it about. I wanted to know what it is that makes people who have a rich and complex culture fall for dry abstractions and mutilate their traditions, their thought, their language so that they can accommodate the abstractions. I wanted to know how intellectuals manage to get away with murder—for it is murder, murder of mind and cultures that is committed year in, year out at schools, universities, educational missions in foreign countries.[43]

For Deloria, this kind of critique made by Western thinkers is extremely important for non-Western traditions because it opens up spaces that are normally closed. Only once the pretension of superior knowledge in Western thought is disabused will Native American even be given a chance to sit at the table and participate in the discussion.

If non-Western thinkers are allowed at the table they might say something like the Osage chief Big Soldier did in 1820:

> I see and admire your manner of living, your good warm houses, your extensive fields of corn, your gardens, your cows, oxen, workhouses, wagons, and a thousand machines that I know not the use of. I see that are able to clothe yourselves, even from weeds and grass. In short, you even do almost what you choose. You are surrounded by slaves. Everything about you is in chains, and you are slaves yourselves. I fear if I should exchange my pursuits for yours, I too should become a slave.[44]

Commenting on this Deloria says: "If we subdue nature, we becomes slaves of the technology by which the task is accomplished and surrender not simply our freedom but also the luxury of reflection about our experience that a natural relationship with the world has given us."[45] From the Native American perspective, the reduction of nature to a machine to be investigated and altered according to human desire is woefully unethical in terms of how it treats the world but also self-undermining to humanity because obtaining this mastery over nature requires stripping reality of meaning and spirituality—which are fundamental to human experience. Ostensibly, it is possible to contain and limit this reduction of the world to its physical nature to scientific theory and practice, while recognizing and preserving the meaningful and spiritual aspects of human experience. Historically however, this is not what has happened. The epistemic reduction of reality has been mistaken as an ontological reality: in other words the epistemic task of studying physical reality has turned into the mistaken view that reality IS only the physical. Values, meaning, and spirituality have been outcast from objective reality to subjective preference.

According to Wildcat, "The idea that morality and values cannot be found in nature is one of the single most erroneous notions Western civilization and modern science have produced."[46] Deloria and Wildcat claim the radical difference between the mechanical world presented by science and the meaningful and spiritual world of everyday human experience reflects a metaphysical schizophrenia. Everyday experience as meaningful and spiritual does not fit with the "objective" truths of science. From the Native American perspective, too many Westerns neglect the fullness of their own experience as a result of this metaphysics. Both for human and for the Earth and all the nonhuman people on it, the fallout of this metaphysical schizophrenia is dramatic.

THE PRACTICAL FALLOUT OF WESTERN METAPHYSICS

Western metaphysics says that the world can be broken apart into its component parts. As a result, Western knowledge fragments the world and tries to understand it in pieces. This results in knowledge that is broken and difficult to bring together into some coherent whole. Between specialization of disciplines, the reduction of knowledge to facts, and the increasing speed to the discovery of new facts it is hard to say how these facts come together to make a meaning big picture of what is happening around us. Wildcat thinks there is simply too much information and like drinking from a firehose, we are not sated by this wild flow but inundated and crushed. What we need is not more information but wisdom and coherency to make sense of all this.[47]

Wildcat points out that there is little support in modern institutions for the "emotional and spiritual development of individuals" and that most important questions of meaning and purpose are left out of public education.[48] Instead, education functions as job preparation and the accumulation of facts.

Existential and spiritual questions are left up to each individual to figure out on their own if they want.

This produces a kind of thoughtlessness as the primary questions for each individual have become "what do I need to do to get ahead?" and "what are the best options for me in this system?" instead of more fundamental questions like "what is ethical?," "what is my purpose in life?," and "why are things this way instead of some other way?" Students are not taught to ask these existential and spiritual questions according to Deloria because "the goal of much of modern education seems to be socialization. That is to say . . . we are training people to present an acceptable profile to the corporate industrial world."[49]

Furthermore, by making questions of meaning and spirituality a matter of individual preference and individual decision, this obscures the ways that we are connected. The result is not just alienation from the Earth, place, and the plants and animals in a place, but from other humans as well. Living in a mechanical, fragmented world results in the modern individual who is "alienated, surrounded by people yet alone, and has a closer relationship with technology than with family."[50]

RECOVERING COMMUNITY AND AMOR MUNDI

From the Native American perspective of Deloria and Wildcat, it does not have to be this way. Traditional Native American communities were taught and recognized their connection to the world around them, fostered relationships with plants and animals and with each other, and found meaning in community. Wildcat says that it has become trendy to reject the idea that Native Americans are some sort of saints in the wilderness because figures like Rousseau manipulated these images to critique Europe, but warns against throwing the baby out with the bathwater: based on the records of anthropologists and colonizers there is actually good reason to take seriously the "generosity and social-well-being" of many Native American communities.[51] Wildcat quotes Christopher Columbus's journal: " 'They are very gentle and without knowledge of what is evil nor do they murder or steal,' and 'They love their neighbors as themselves': not bad accomplishments for 'savage heathens.' Unfortunately, these qualities were not much valued by Colon and many of the Europeans who followed him."[52]

Again, Wildcat is not trying to claim all Native Americans are saints. Romantic thinkers like Rousseau famously portrayed Native Americans as "noble savages" and used them as a foil to critique Western culture. A common strategy to push back against this kind of romanticism is to go the other way and not only seek to undermine these romanticized images of premodern peoples, but often stress how miserable everyone was previous to modern science and modern economics. Especially in economics, it is common to portray precapitalism as a disastrous, backward time in which everyone was near dead from diseases and lack of food. This is the sort of thing that authors like Marshall Sahlins and Lewis Hyde are trying to push back against.[53] Sahlins writes a book he titles *Stone Age Economics* in which he argues against modern economism that premodern people not only often lived comfortably but also did not have many of our modern neuroses and alienations.

Against this anti-romantic dismissal of premodern peoples, Wildcat claims that "it is not romanticism to suggest that Seattle, Ten Bears, Chief Joseph, and many other American Indian leaders of the nineteenth century lived in environments where the notion of a 'struggle for existence' never crossed their mind—although concern for living well did."[54] In other words, Wildcat claims that it is not wild fancy to take seriously the possibility that Native Americans had healthier, less environmentally destructive practices than the West. In fact, Deloria argues, if we pay attention to the accumulating evidence from within Western science itself, then we should recognize that Native American metaphysics is a better fit with this new evidence than schizophrenic Western metaphysics.[55]

In terms of *amor mundi*, it seems that Native American philosophy and practice does support an approach to the Earth and Nature that can be characterized as a love of the world. As we are warned by Deloria, we have to understand this the right way to make sure we are not mischaracterizing Native American experience. When Deloria claims that Native Americans do not love nature, he is rejecting the tendency in Western thought to work in terms of abstractions. Thus, his claim is that there is no abstract thing "nature" in Native American thought, just as there is no such thing as "spirituality" or "religion." Rather, Native American experience as a whole is suffused with spirituality, religion, and nature in a way that makes it incomprehensible to conceive of them as separable from experience.

So when Robin Wall Kimmerer attempts to teach us to relate to sweetgrass, she is not seeking to help us relate to the abstraction "nature" or to an abstraction "sweetgrass" but to this particular sweetgrass she is holding. Abstract categories are too vague. Deloria claims that Native American metaphysics implies we must related to sweetgrass as a person, in a personal way. The sweetgrass here, in this location, will be different from the sweetgrass on the other side of the mountain. They may be similar, but they may grow in different soil, with different predators, different light, different water, and so on. To treat all of these specific persons as a part of some abstract category "nature" and then to claim to love Nature or the Earth risks completely losing the particular in the abstraction.

However, I think we can conclude that there is a powerful love of the Earth (understood as personal relationships with the specific persons in a particular place) in Native American thought. As I mentioned earlier, of the three kinds of *amor mundi* I have been examining, this is the kind of love that is most easy to experience and recover. This is not to say that a loving commitment to a place is easy—in fact, for most of us living in modern consumer society it requires radical changes in our everyday practices and thinking. But there are many environmentally minded thinkers who are already attempting to change their lives, to change society, and to challenge the dominant paradigm.

Speaking on a personal level, after reading Deloria and Wildcat, I was moved to try and take place more seriously. I began to earnestly learn about the landscape, animals, plants, and history of the place I was living in. It was not that I was not interested in these things before, but rather I began to think of learning about the history and nature of an area as part of an ethical task of relating to it. I have found that this has altered my perception of the world. It has even retroactively changed my relationship to places I have lived in the past. I am not claiming to experience grass or moss on the level described by Robin Wall Kimmerer. But I am claiming that with the right guides, we can open ourselves up to the powerful experience of loving the Earth and Nature in a way that will fundamentally change us. I will point to their works of some of these guides in my conclusions.

But first I want to turn to the question of modern atomism and the moral forces that have made it so pervasive and "natural." Any recovery of *amor mundi* will require overcoming modern atomism, but this turns out to be a complicated task.

NOTES

1. His approach in "The Metaphysics of Modern Existence" is to claim that there is increasing recognition in Western thought and science of the flaws of traditional Western science.
2. Sturgeon, Noël. "Naturalizing Race: Indigenous Women and White Goddesses." in Noël Sturgeon, ed., *Ecofeminist Natures* (New York: Routledge, 1997).
3. Deloria, Jr., Vine and Daniel Wildcat. *Power and Place: Indian Education in America* (Golden, CO: Fulcrum, 2001), 5.
4. Deloria, Jr., Vine. "Kinship with the World." *Journal of Current Social Issues*, Vol. 15, No. 3, Fall 1978, pp. 19–21.
5. Deloria, "Kinship with the World."
6. Ibid.
7. Ibid.
8. Ibid.
9. Deloria and Wildcat, *Power and Place*, 5.

10. See the previous two chapters.
11. Ibid., 11.
12. Ibid., 12.
13. Ibid., 22–23.
14. Ibid., 60.
15. Ibid., 24.
16. Ibid., 62.
17. Ibid., 64.
18. Ibid., 63.
19. Wall Kimmerer, Robin. *Braiding Sweetgrass: Indigenous Wisdom, Scientific Knowledge and the Teaching of Plants* (Minneapolis: Milkweed Editions, 2015), ix.
20. Wall Kimmerer, *Braiding Sweetgrass*, ix–x.
21. Deloria and Wildcat, *Power and Place*, 12–13.
22. Wall Kimmerer, Robin. *Gathering Moss: A Natural and Cultural History of Mosses* (Corvallis, OR: Oregon State University Press, 2003),
23. Atleo, E. Richard. *Tsawalk: A Nuu-chah-nulth Worldview* (Vancouver: UBC Press, 2004).
24. Atleo, *Tsawalk*, 61.
25. Ibid., 63.
26. Ibid., 12.
27. Tinker, George. "Religion" in Frederick Hoxie, ed., *Encyclopedia of North American Indians* (New York: Houghton, 1996).
28. This theme is present in writings of all the Native American authors mentioned in this chapter.
29. LaDuke, Winona. *Recovering the Sacred: The Power of Naming and Claiming* (Chicago: Haymarket Books, 2016).
30. Deloria, Jr., Vine. *Spirit and Reason: The Vine Deloria, Jr., Reader* (Golden, CO: Fulcrum, 1999), 229.
31. Deloria, *Spirit and Reason*, 225. See Jaspers, Bellah and Taylor also.
32. Ibid., 225.
33. Ibid., 225.
34. Ibid., 225–226.
35. Deloria and Wildcat, *Power and Place*, 12.
36. I discussed this issue at the end of Chapter 2.
37. Deloria, Jr.,Vine. *The Metaphysic of Modern Existence* (New York: Harper, 1979), ix.
38. Pack, Justin. *How the Neoliberalization of Academia Leads to Thoughtlessness: Arendt and the Modern University* (New York: Lexington Books, 2018).
39. Deloria, *The Metaphysic of Modern Existence*, xi.
40. Deloria, *Spirit and Reason*, 3.
41. Ibid., 5.
42. Feyerabend, Paul. "How to Defend Society Against Science." *Radical Philosophy*, Vol. 11, No. 1, 1975, pp. 3–9.
43. Feyerabend, Paul. *Against Method* (New York: Verso, 2010), 278.
44. Deloria, *Spirit and Reason*, 4.
45. Ibid., 4.

46. Deloria and Wildcat, *Power and Place*, 136.
47. Ibid., 31–32. There is fantastic quote I would love to include here, but the publisher wants a hefty fee in order for me to do so. So we can add to the problems with too much information the monetization of it.
48. Ibid., 32.
49. Ibid., 79.
50. Richard, *Tsawalk*, 34.
51. Deloria and Wildcat, *Power and Place*, 141.
52. Ibid., 142.
53. Sahlins, Marshall. *Stone Age Economics* (Chicago: Aldine, 1994); Hyde, Lewis. *The Gift: Imagination and the Erotic Life of Property* (New York, NY: Vintage, 1983).
54. Deloria and Wildcat, *Power and Place*, 141.
55. Deloria, *The Metaphysic of Modern Existence*.

Chapter 6

Overcoming the Modern Juggernaut

WHERE DOES THIS LEAVE US?

If the arguments we have seen so far are correct then the cosmos is dead and we have killed it. We now live under the "ontological dominion of death" as Jonas calls it and largely no longer experience our world as a living, meaningful place.[1] As we saw in chapter 2, the disenchantment of the cosmos has resulted in a mechanical universe which part of our modern morality—technology—calls us to understand and "fix" where necessary.[2] This relationship is not about love but control and mastery. At its worst, it even seems to be based on resentment and hatred. Furthermore, the necessity of objectivity to discover the laws of the universe supposedly requires the rejection and control of emotions, including love, and the reduction of reality to only the physical. As a result, we find ourselves in a universe which has no given morality or inherent meaning (these are now in the human mind), which has no teleological purpose and which we can manipulate and fix through objective, rational control. Love is excluded from science and rationality and is no longer located in the world. Modern conceptions of love make it a subjective emotion, powerful, but increasingly individualized, eroticized, and commodified. The premodern experiences of love of the cosmos, community, Earth, and nature are largely forgotten and, with occasional exceptions, generally foreign in modernity.

The disembedding and buffering of the modern self has disconnected us not only from the cosmos but also, as we saw in the last three chapters, alienated us from the Earth, Nature, and traditional, interconnected human communities. This has led to massive environmental damage and the systematic destruction of most traditional communities. The project of modernity has sought not only to discover the laws of the universe to make the Earth more available for human use but has also turned humanity itself into a resource to be managed and improved.[3] This has led to the modern forms of labor alienation analyzed by Marx and the myriad critiques marshaled by the Romantics and other critics of modernity.[4] After being disembedded and buffered from the cosmos, the Earth, nature, and traditional human community, we end up with the hyper-individualism of modern consumer society. In the end, do we end up with the isolation of the lonely crowd?[5] Solipsism? The nihilism of the last man? As worlds multiply and the hold of any particular world narrows and shrinks, is it any surprise that the project of modernity has led to the decline and forgetting of *amor mundi*?

Now that we have examined some different kinds of *amor mundi*, in this chapter, I want to think about why we need it more than ever. I will claim learning to love the world is an existential, social, and environmental task. We need love. But we need more than modern anthropocentric, commodified love. We need to recognize our connection and dependence on the Earth, on nature, and on our communities. We need to feel connected to the past and committed together to our future. We need to reject and overcome the false modern understanding of the self as an autonomous individual that we examined in chapter 2. This atomistic view of the self is tied to a complex set of assumptions about rationality, gender, race, emotion, and so on, and has become the (bad) foundation of the primary economic and political theories of modernity.

I want to return to the discussion of the modern self from chapter 2 and tie it together to the critiques offered by Arendt and Deloria and Wildcat. I will begin by focusing on modern love, and how it has become drastically narrowed to something almost solipsistic. To understand this shift, I will turn to the modern understanding of the self and discuss not only why it is critically flawed, but the moral attraction of the modern understanding of the self. Understanding the moral power of the modern self is particularly important to explain why, despite powerful arguments against it, the modern understanding of the self retains such a hold on the modern moral imaginary. Only once the powerful hold of this modern understanding of the self is broken, can we begin healing our fragmented relationships with the Earth, nature, and each other.

A VERY BRIEF HISTORY OF LOVE IN WESTERN PHILOSOPHY

Let me offer a brief overview of the history of love in Western philosophy to show the gradual narrowing of the scope of love and its modern sexualization. As we saw in chapter 1, Plato and much of the ancient world not only had an ambivalent attitude toward sex, the physical body, emotion, and love but also embraced contemplation (of the cosmos and later God) as the supreme human activity. I have claimed that contemplation itself was a kind of love: the establishment of a deep connection and relationship with the cosmos or God. Irving Singer, in his classic three-volume history of love in Western philosophy claims that there were two main strands of this tradition. The first, following Plato but also especially Plotinus, emphasized *eros* and merging. The second, following Aristotle, emphasized *philia* and union. The Platonic *eros* tradition sought mystical merging with the cosmos or God—connecting with it so deeply that one merged with it in some way. The Aristotelian *philia* tradition sought not a merging, but a union in which those involved remained separate but accomplished some degree of unity.[6]

In terms of scale, in the Platonic, cosmic *amor mundi* the connection and relationship that love established was immense. The Aristotelian *amor mundi* is smaller in scale and more often associated with the kind of political, friendly commitment, and love examined by Arendt. Singer claims that medieval Christianity retains to some degree these Greek commitments to the cosmos and community, but the fundamental concern shifted to God. Furthermore, Singer claims that medieval Christianity combined these Greek forms of love with Hebrew forms of love, *nomos* and *agape*. On Singer's reading, *nomos* refers to the love that manifests as obedience to God's will and *agape* refers to God's love for humanity.

While the shift to Christianity can, in some figures (most famously Francis of Assisi), still manifest a deep love of the cosmos, in general the Christian emphasis tends to be anthropocentric. On Brague's reading of Plato, humans do have a special relation to the cosmos, but the cosmos is not created for humans by a creator God like the Biblical tradition.[7] In the Biblical traditions, humanity has privilege of place over the rest of creation as the culminating creation of God, and this is reflected in Biblical ethics, for example, when Jesus makes it clear that the two great commandments are to love God and to love our neighbors. This contrasts with the common Native American account we saw in the last chapter which also has humans as the last creation, but concludes that humans are the younger siblings of the rest of creation. Moral responsibilities here extend to all the peoples (plants, animals, rocks, rivers, etc.) of a particular place. From the Native American perspective, Western thought manifests a disturbing anthropocentric narrowness.

The medieval Christian period sees a shift toward a more anthropocentric understanding of love that will only continue with the rise of modernity. But we also find in this period an attempt in courtly love to overcome the Greek and Christian ambivalence about sex.[8] Thus, interestingly, Singer points out that while medieval theologians focused on mystical merging with God (*eros*), brotherly love (*philia*), and the love of God for humankind (*agape*), medieval courtly love came to reject the Greek and Christian devaluation of sex and sought ways to cultivate sexual intimacy. While some opponents attacked this "as a cunning glorification of sex," Singer claims that sex was treated as "subordinate to the quest for values that effect an honorific union" but a potentially important aspect of noble relations.[9]

This revalorization of sexual love was greatly magnified by the Romantics. Romantic love interpreted "beauty or goodness in terms of the erotic experience itself" and "elevates sexuality into something super-physical, metaphysical, something transcendental and more than merely biological."[10] Here the powerful tensions of sex were valued as heightening human experience, enabling it to transcend the mundanity of human experience. Art, poetry, and literature are all sensory in a fundamental sense, and Romanticism embraced the sensual, even sexual richness of life. This heightened richness obtains an odd salvific power. Just as sex is more intense than normal life, so too the Romantic eroticization of experience and beauty sought to rescue life itself from the monotony of the growing Industrial Revolution.

The twentieth century continued the sexualization of love in complicated ways that I cannot fully delve into here.[11] But at a general level one of the largest factors is the rise of modern advertising, which, in a certain sense, strips away much of the spiritual aspects of the Romantic valorization of sexuality and leaves us with raw sexual desire as means to sell products. The event that accelerated the use of advertising was the invention of the modern assembly line. As the assembly line became widespread, companies had to deal with the problem which Stewert Ewen calls the "social crisis of production": how to sell their new glut of products to a populace that had traditionally valued frugality.[12] Some advertising had existed previously to the assembly line, of course, but now it was marshaled not just to make people aware of products but also to create and stimulate artificial needs.[13] Frugal Americans needed to be fundamentally transformed from producers and savers to consumers and spenders. Ads became more complicated, playing on social status, creating insecurities, and promising solutions through smart shopping. Since this often meant creating and stimulating desire, sexualized imagery became more prominent. Sex sells, and as consumers were increasingly bombarded with sexualized imagery, love itself became increasingly associated with sex.

While medieval courtly love, Romanticism and modern advertising eroticized sex, the Renaissance, Reformation, and Enlightenment continued the well-documented shift toward modern individualism that narrowed the scope of love further from the grand scale of the cosmos and community through a series of disembeddings and the buffering of the modern self until we arrive at the familiar atomistic individualism of modernity.[14] When combined with the sexualization of love, we find the common contemporary understanding of my students, who tend to primarily articulate love in terms of sexual desire between partners.

ANTHROPOCENTRIC, ATOMISTIC, SEXUALIZED LOVE

The history of love in Western thought shows that love has become increasingly anthropocentric, individualized, and sexualized.[15] I have presented a very brief overview of the history of love in Western thought to outline the general trajectories of these shifts.[16] The current culmination of this history is that contemporary love is often primarily articulated in terms of romantic and/or sexual love between individuals. We can see this individualism not only through comparative and historical analysis but even in the kinds of topics that are addressed in academic venues that are centered on love. Take, for example, the discussion in the relatively new Society for the Philosophy of Love and Sex, which has existed since 1977. Organized by Alan Soble, a key modern popularizer of these issues in philosophy and a prolific author in his own right, the first volume of essays from this organization is described thus:

> This collection joins together sixty essays on the philosophy of love and sex. Each was presented at a meeting of The Society for the Philosophy of Sex and Love held between 1977 and 1992 and later revised for this edition. Topics addressed include ethical and political issues (AIDS, abortion, homosexual rights, and pornography), conceptual matters (the nature, essence, or definition of love, friendship, sexual desire, and perversion); the study of classical and historical figures (Plato, Aristotle, Spinoza, Kant, and Kierkegaard); and issues in feminist theory (sexual objectification, the social construction of female sexuality, reproductive and marital arrangements).[17]

The second volume, which includes essays from 1993 to 2003 is described as follows:

> Some of the best minds on three continents, from four nations, and eighteen of the United States discuss such topics as adultery, commitment, cross dressing, gender politics, date rape, family, friendship, friends as lovers, gayness, love, marital pluralism, marriage, prostitution, religiously motivated anti-queer sentiments, same sex marriage, seduction, and self-respect.[18]

These topics are all important. I quote these descriptions not to diminish these issues, but rather to point out that *amor mundi* in some form or another is largely if not wholly absent. Most of these topics have to do with love and sexual relations between individuals or social attitudes about them. In short, the focus here is anthropocentric and individualistic love. Occasionally, there is an analysis of love in abstract or more general terms, but the pieces that go beyond individual love still tend to be anthropocentric and don't tend to analyze the kind of worldly loves we have been analyzing in this book.

A similar pattern emerges in philosophy of love and sex courses. An unscientific perusal of online syllabi shows that philosophy of love and sex syllabi tend to present some history of love and sex and primarily deal with the ethics of interpersonal relations, especially sexual relations. This is not surprising considering that the primary audience—young college students—are often intensely interested in the subject matter. As such, discussing the philosophical issues of love and sex is a good way to show students the power of philosophical analysis and the relevance of philosophy. Unfortunately, perhaps because of this interest, philosophy of love and sex courses are often lacking with regard to nonsexual relations of love—like the love between parents and children or between friends. The kinds of love of the world that we have been examining are basically never present.

There are good reasons to worry about atomistic love. When the sexualization of love is combined with modern atomistic individualism the result is a disappearing point of relation-less masturbation. And indeed, it looks like this is where we are headed: in a recent article Atlantic Monthly writer Kate Julian argues that "despite the easing of taboos and the rise of hookup apps, Americans are in a sex recession."[19] Young Americans are actually having less sex than their parents did at the same age, who in turn have/had less sex then their parents did at the same age. This is not just a pattern among Americans, it is happening in a variety of first world countries including Japan, Britain, Australia, Finland, and so on. At the same time, rates of porn use and masturbation are on the rise (by both men and women). While this "sexual recession" is a new phenomenon and a complicated one, Julian's basic claim is that online porn and masturbation is easier than dealing with a real person, so young people, despite more sexual opportunity, are actually becoming less likely to have intercourse and instead find sexual "fulfillment" in masturbation. The point is not just about sex, but about the increasing tendency to withdraw inwards and lose connections with others in general.

If Julian is right, this would be the disturbing and poetic culmination of the gradual modern disembedding and buffering of the self. Nothing is left but isolated selves, maximizing their pleasure in digital worlds. The film *Children of Men* imagines a desperate future where humanity loses the capability to have children. Perhaps a more likely scenario would be one in which no one can be bothered to have children—or even sexual intercourse with each other. Perhaps a more accurate film is Spike Jonze's film *Her*, in which the lonely protagonist falls in love with an operating system.

This modern shift away from community and world toward the atomized self has been tracked, documented, and critiqued by social critics, especially sociologists. From the early concerns of Ferdinand Tönnies and Emile Durkheim,[20] the midcentury warnings about the "lonely crowd"[21] and the "society of individuals,[22] to contemporary accounts of postmodern individualism,[23] sociologists have tracked the rise of this radical individualism and worried about the fallout. Twentieth-century German philosophers have been attentive to these critiques also. In general, mainstream Anglo-American philosophy, however, has not. Neither have American economists and political scientists. In these disciplines, the modern atomistic self is largely taken for granted, despite good philosophical reasons to reject it. Why, despite these powerful critiques of this understanding of the modern self, has it become so foundational to modern economics and politics?

THE MODERN SELF: WHY THIS POOR UNDERSTANDING OF THE SELF REMAINS SO INFLUENTIAL

I discussed this in some detail in chapter 2, but we are now in a good position to return to it. I want to begin this by focusing primarily on the work of Charles Taylor. This is for two reasons. First, the topic is sprawling and can easily spin out of control. Using Taylor to ground the discussion should help limit this. Second, Taylor shows how disparate and seemingly unrelated concepts and areas of study are often connected by common moral threads that serve both as tacit assumptions and tacit moral ideals. Thus, the epistemology of a particular tradition may share certain tacit moral evaluations (say, that a "scientific" and "rational" approach is better than a poetic one) with the economics or the cultural understanding of human nature of this same tradition. Taylor's claim is that there is a complicated, interconnected set of ideas that reinforce each other across diverse topics that have become widely taken for granted in modernity despite being critically flawed. Taylor has made the question of "overcoming" this modern "Hydra whose serpentine heads wreak havoc throughout the intellectual culture of modernity—in science, in criticism, in ethics, in political thinking, almost anywhere you look" the central focus of his life's work.[24] Again, although each "head" of the hydra is often presented as separate (supposedly epistemology, ethics, politics, the nature of the self, social theory, philosophy of language, and so forth, are independent), they actually draw on common moral ideals and sources and stealthily strengthen each other. While each of these interconnected ideas have been systematically challenged and critiqued, they continue to be widely influential in modern thought and modern society precisely because of these tacit moral sources and ideals.

When I claim that these moral sources are "tacit" I do not mean that modern philosophers and scientists are actively hiding them, but that these moral sources and ideals are often driving the arguments of modern thinkers without their being aware of it. Let me give some examples: without arguing why, most modern thinkers take for granted that we should be clear, transparent, and organized. These are often uncritically taken to be good, while fuzzy, murky, and disorganized are bad. Logic and "rationality" bring about clarity, so they must be good. As we saw in chapter 1, emotion, love, and poetry are often portrayed as difficult and confusing. Clarity, transparency, and organization have/gain a moral force that becomes present in many subtle ways in our speech, our actions, and our relationships.

It may be tempting to say: who doesn't want clarity? In *The Genealogy of Morals*, Nietzsche points out that "Truth" is one of these taken-for-granted moral valuations of many traditions.[25] But, Nietzsche asks, why should we want the Truth so much? What if a joyous life is built on lies? Is that bad? Should we eliminate falsehood if it brings happiness? Similarly, what if clarity or organization made us miserable? There are, Nietzsche claims, widespread, taken-for-granted moral assumptions or ideals at work in any tradition—if these are challenged the response is often confusion ("Why in the world would someone not want the truth?"). Part of the power of the comparative and historical genealogical approach that Nietzsche wielded is to dig up these assumptions and question them.

Since these moral valuations are buried deep, embedded in our language, our thought and our actions, they often function "behind our backs." In *Metaphors We Live By*, Lakoff and Johnson show how our language is completely suffused with this kind of moral valuation: front, forward, and up are generally good, back, backward, and down are generally bad.[26] One goes up to heaven and down to hell. A good society is moving forward and not backward. They point out how money metaphors have become common in English: "Don't waste your time." "Spend your time wisely." "I'm buying in." "She is invested in the project." The incorporation of money metaphors into everyday phrases suggests how profoundly money is structuring how we understand our experiences. Thus, the centrality of economics to modern human life is reflected in the increasing ubiquity of economic language in our everyday language.

This is not to say that all the moral sources and ideals at work in our lives are in harmony or that they are a monolithic set. On the contrary, there are often contradictions and tensions at play. In Taylor's *Sources of the Self* and *A Secular Age*, he attempts to historically trace some of the key conflicts and tensions in modern thought. These are massive, learned texts, and I have already discussed Taylor's account of the disenchantment of the cosmos, the gradual disembedding of the self from the world that replaced the premodern porous self with the modern buffered self in chapters 1 and 2. My focus here will be on the moral sources and ideals that drive this shift and bolster our modern valorization of individualism.

Taylor claims that there are three fundamental foci of modernity that are interconnected, widely influential, and fundamental to modern political and social theory and our modern understanding of the self: individualism, instrumental rationality, and freedom. Taylor describes how they come together to promote a certain vision of the individual:

> The first [connected notion] is the picture of the subject as ideally disengaged, that is, as free and rational to the extent that he has fully distinguished himself from the natural and social worlds, so that his identity is no longer to be defined in terms of what lies outside him in these worlds. The second, which flows from this, is a punctual view of the self, ideally ready as free and rational to treat these world—and even some of the features of his own character—instrumentally, as subject to change and reorganizing in order to better secure the welfare of himself and others. The third is the social consequence of the first two: an atomistic construal of society as constituted by or ultimately to be explained in terms of, individual purposes.[27]

Simply put, modern individualism is associated with rationality, freedom, self-reliance, autonomy, independence, and control. This image of the modern individual functions on multiple levels. On the surface it functions supposedly as an "objective" description of the modern individual, but Taylor claims this masks that it is actually NOT a very good description of what humans are like and reflects instead a certain moral image of what modern philosophers thought humans SHOULD be like. In other words, in practice humans are not as free, nor as rational, nor as atomistic as this description. If this is the case, why do these connected notions remain so influential? Taylor claims moral ideals about self-responsibility, freedom, autonomy, and how both an individual and society in general should function according to these ideals influence the enthusiasm for this modern understanding of the self. As we saw in chapter 2, what is being established and invoked in modernity is a model of a civilized, independent, rational man that is the building block of a civilized, rational (productive and efficient) society.

This (idealized) image of a human as free, autonomous, and rational is one that is key to modern economic and political theories. Modern economics and politics tend to begin with the assumption that society is a collection of individuals who exist first or primarily as individuals but who come together to accomplish things as a society that they cannot on their own.[28] The individual is taken to have certain preferences and to act rationally (i.e., calculatively or instrumentally) in order to bring about results that they want. Individuals come together to accomplish things that they cannot do by themselves, giving up some of their freedoms in the process. The atomistic individual is taken as the fundamental building block and the unit of analysis for understanding what is meant by freedom and rationality.

This should sound like a common approach in modern science—analyzing a particular phenomenon in terms of what are taken to be the constituent parts and how they interact. Taylor says we see this "atomistic" approach being applied widely in the social science and humanities. Take, for example, modern theories of language.[29] In many cases, we see a focus on individual words as the fundamental building blocks of language. Taylor points to Condorcet who describes two humans meeting in a desert and attempting to communicate. They do so by pointing at something and pronouncing their word for it. Bit by bit, by explaining one word, and then another, they build up to understanding each other's languages. But both twentieth-century analytic philosophy and twentieth-century continental philosophy have shown that this is not how language functions. If anything, each word only makes sense in relation to the whole.[30]

Similarly, humans do not exist by themselves first in a desert and then come together to form a community. We are born into a community and raised by parents or other people. If anything, the community comes first and the individual comes later. This is a point hammered home by Deloria and Wildcat. Furthermore, the image of humans being instrumentally rational is not correct. Humans do many things for many reasons. Narrowing all this complexity down to the instrumental calculation of maximizing preferences diminishes it. And yet economics is largely based on both this individualism and the supposed instrumentality of the modern individual.[31]

Modern individualism is not then something that was discovered, some fundamental truth about humans, but a particular ideal of what humans can be like—something that humans can be made into: *homo economicus*. For thinkers such as Foucault, Bauman, and Castro Gomez, the modern period is marked by the project to remake humans into disciplined, "rational," "civilized," productive self-monitoring individuals.[32] This ideal and the ensuing disciplinary project has been and continues to be highly contested. Taylor calls the fight over these conceptions of individualism, instrumental rationality, and freedom the malaises of modernity. He means by this that as modern society has increasingly transitioned toward these particular ideals, they have become sources of worry and intense debate, both by those who oppose them but also by those who think we are failing to live up to them.[33] It is worth quoting Taylor at length about these tensions with regard to modern individualism:

> Of course, individualism also names what many people consider the finest achievement of modern civilization. We live in a world where people have a

> right to choose for themselves their own pattern of life, to decide in conscience what convictions to espouse, to determine the shape of their lives in a whole host of ways that their ancestors couldn't control. And these rights are generally defended by our legal systems. In principle, people are no longer sacrificed to the demand of supposedly sacred orders that transcend them.
>
> Very few people want to go back on this achievement. Indeed, many think that it is still incomplete, that economic arrangements, or patterns of family life, or traditions notions of hierarchy still restrict too much our freedom to be ourselves. But many of us are also ambivalent. Modern freedom was won by breaking lose from older moral horizons. People used to see themselves as part of a larger order. In some cases, this was a cosmic order, a "great chain of Being," in which humans figured in their proper place along with angels, heavenly bodies, and our fellow earthly creatures. Their hierarchical order in the universe was reflected in the hierarches of human society. People were often locked into a given place, a role and station that was properly theirs and from which it was almost unthinkable to deviate. Modern freedom came about through the discrediting of such orders.
>
> But at the same time as they restricted us, these orders gave meaning to the world and to the activities of social life. The things that surround us were not just potential raw materials or instruments for our projects, but they had the significance given them by their place in the chain of being. The eagle was not just another bird, but the king of a whole domain of animal life. By the same token, the rituals and norms of society had more than merely instrumental significance. The discrediting of these orders has been called the 'disenchantment' of the world. With it, things lost some of their magic.[34]

Similar tensions are involved with regard to instrumental rationality and freedom. Critics have worried that instrumental rationality is predatory and inhuman.[35] De Toqueville famously outlined the dangers of American democracy and freedom.[36] These are highly contested and interrelated pillars of modern social, economic, and political theory. To reject them seems to reject modernity itself.

The power of Taylor's argument is his claim that the moral function of these issues has been elided. In order to look scientific, social, political, and economic theories are often presented without reference to morality. They are treated less as theories of how things should be and more as descriptions of how things are. Their lingering hold over the modern imaginary then is that these ideals have established themselves as supposedly scientific (which they are not) while simultaneously exercising a hidden (nonscientific) moral force. Modern atomism both IS the reality but also SHOULD be the reality. This means that advocates of modern society (economists for example) can claim both scientific authority AND moral authority. If you reject the moral force, they can claim they are just working with the data. If you question the data, they can call you backward.

This is only part of the story. As we also saw in chapter 2, the issue is still more complicated than this because the modern, free, civilized, rational, disciplined, independent self has been and often still is racialized and gendered. In other words, modern man was defined over and against primitive, superstitious nonwhite peoples and emotional, irrational women[37]. To add insult to injury, the supposed inferiority of primitive people was used as a justification to colonize and sometimes forcibly modernize them.[38] The supposed inferiority of women was used as a justification to exclude them from the public and treat them as second class citizens.[39]

Unfortunately, there is a widespread but mistaken tendency to assume that once colonialism is ended or women get the vote or segregation is ended that these issues will be over.[40] But in light of Taylor's discussion of how these moral sources and ideals tend to function tacitly in our practices, thinking, and language, it should be clear that the end of colonialism cannot be the end of colonialism until all the underlying colonial assumptions and ideals are eliminated. But this can be a very difficult and long process if these ideas are deeply rooted in language and practice.

This problem of entrenched modern moral assumptions and ideals affects the problem of recognizing and overcoming world alienation and it is reflected in our inability to do anything about the critical environmental problems we are facing. Despite widespread awareness of the environmental crisis, seemingly nothing changes. While there are various factors that contribute to this, it is clear that one of the fundamental problems is our hyper-individualism and our disconnection with nature and the Earth. Of the Gordian knot of modern assumptions and ideals there is one thread that is most pertinent to the question to *amor mundi*: the modern moral assumption and ideal of independence. I want to turn to this particular issue.

THE MORAL IDEAL OF INDEPENDENCE

Modern western morality (especially in the United States) highly values independence. We are taught that we should be independent thinkers, financially independent, emotionally independent, self-reliant, and autonomous. Along with this comes an often intense moral opprobrium on dependency: we should not live with our parents too long, mooch off others, depend on the government for welfare, fail to stand on our own feet, and so on.

According to Nancy Fraser and Linda Gordon, this contemporary understanding of independence and dependence is actually rather new. The preindustrial English understanding of dependence related it to subordination, but not necessarily as a negative.[41] Religiously, all were supposed to be dependent on God. If anything dependency was normal. Independence was first applied to individual humans in the eighteenth century to describe those who had enough money that they did not have to labor. This sense of independence is still used recognizably in English when someone is described as "independently wealthy."

As Arendt points out in *The Human Condition*, democratic citizenship and participation in the Greek polis was originally only allowed for those of independent means.[42] This was because only those who were free from labor could have the time and the appropriate disposition to participate in the polis. The worry was that someone primarily engaged in work or labor would approach the polis with the same mentality that guided their work or labor, but the polis was a place that functioned outside of the logic of work and labor and had to be protected from them in order to preserve its freedom. A similar attitude can be seen in the early years of democracy in the United States: only white, landowning males were allowed to vote. Ostensibly, only those with sufficient time and education would be able to vote competently and so the vote was limited to these individuals.

The vast majority of people in America—most white men, all women, and all nonwhites—were not allowed to vote. While they were limited by their dependency, this

> was a normal, as opposed to a deviant, condition, a social relation, as opposed to an individual, trait. Thus, it did not carry any moral opprobrium. Neither English nor U.S. dictionaries report any pejorative uses of the term before the early twentieth century. In fact, some leading preindustrial definitions were explicitly positive, implying trusting, relying on, counting on another, the predecessors of today's *dependable*.[43]

According to Fraser and Gordon, this changed in the 1800s. Independence and dependence featured prominently with industrialization. The language and moral values of radical Protestantism, which Weber famously called "the Protestant ethic" came into increasing prominence and both emphasized manly independence, hard work, and self-reliance. White males wanted both to accomplish these ideals but also to participate in the freedom or citizenship that was allowed to the independently wealthy and free landowners. The problem was that industrial workers could not really claim to be independent. They were clearly dependent because they worked for others. Furthermore, they were often not educated.

In order to be considered "independent" despite their condition of dependency, these terms had to be modified. Fraser and Gordon claim the invention of the family household and family income allowed for this shift in terminology. Independence was loosened up and applied to those men who were "heads of household" and earned a "household wage."[44] Dependency became an increasingly negative category and was affixed to three groups: the pauper (those so poor they relied on poor relief), slaves, and housewives. Additionally, dependency was increasingly no longer considered a social relation, but a character trait of an individual, an entire race or an entire sex (women). Even as women and African Americans won the right to vote, the stigma of dependency sometimes remains and underlies sexist and racist stereotypes to this day.

The problem with this moral valorization of independence and opprobrium on dependence is that no one is truly independent. Even the most obscenely wealthy capitalist was born, raised, educated, and thus dependent on others when they were young. They continue to depend on others to grow their food, make their clothes, build and upkeep their houses, pave their roads, fix their cars, and so on. Furthermore, they will someday grow old and need care. Only some sort of Robinson Crusoe is independent from other humans and even then he is reliant on nature and the Earth (as Bonnie Mann has pointed out).[45]

Eva Kittay claims there is something damningly fantastical about this ideal of independent individuals. Like Taylor, she claims that modern economic and political theories often begin with autonomous individuals and conceive of society as the coming together and organization (social contract) of these individuals into some larger group that provides mutual benefits (she points out the oddness of Hobbes treating individuals as if they had "sprung out of the earth, and suddenly, like mushrooms").[46] On these models, the individual is the fundamental building block of society and rights are framed in terms of individual rights. Freedom is a matter of individual choice. For Kittay, this is a deeply mistaken understanding of both the individual and society. Disturbingly, it forgets that "everyone is some mother's child."[47] By making society the creation of individuals, we completely misunderstand that humans are social beings and that we are born into society, learn the language and customs of a society and are dependent upon it in many ways.

This economic and political "conception of society as an association of equals masks the inevitable dependencies and asymmetries that form part of the human condition—those of children, the aging and the ailing—dependencies that often mark the closest human ties."[48] This forgetting of mothers, children, the disabled, and the elderly is really astonishingly strange. The idea that society is made up of autonomous individuals making instrumental rational decisions based on their preferences is utterly bizarre—not only because of how fantastical this image is but also how out of touch with the basic lived reality around us one must be in order to take such an image seriously. Dorothy Smith says this is the kind of theory that could have only been articulated and supported by theoretically minded (and likely wealthy) men who were fundamentally out of touch with the basic realities taking care of children and other dependents.[49]

The problem is that this is not just a poor description of human life, but a poor moral ideal. Frustratingly, those who buy into this kind of theory then tend to attempt to structure our economic and political institutions based on this model, which often functions both as a supposed description of reality and a moral ideal. This model not only is a bad description of reality but also imposes this image onto the world and thus sees wrongly (actively and selectively misperceives reality): "The presumption of equality obscures the extent to which many of our society interactions are not between persons symmetrically situated, even when they are between individuals who might otherwise be autonomous."[50] The theory functions like a lens that magically disappears dependency, social asymmetry, and the power discrepancies that it entails.

In addition to obscuring the universal reality of dependency, it treats it as a character trait and demonizes it. Fraser and Gordon show how "dependency" is articulated as an individual failure that needs to be rectified by force.[51] While both Republicans and Democrats as late as the 1970s in the United States supported aide for single mothers so that they could raise their children, Republicans have increasingly turned single mothers (especially if they are black) into a boogeyman and supposed drain on society. They have leveraged fearmongering about dependency in order to undermine unions, cut social safety nets, and neoliberalize modern institutions. Dependency in this language is treated as a moral failure and contrasted with the independent entrepreneur. Responsibility in these terms is primarily about whether one makes money, not whether one takes care of children, elderly, or other dependents.

This delusion about independence, autonomy, and self-responsibility has resulted in a widespread political atomism.[52] Kittay asks us to imagine how different it would be if we organized society based on the assumption that the basic building block of society was not the autonomous male—*homo economicus*—of political theory and economics, but a mother (or any caretaker) with a dependent: if that were the case, society should be structured to support caretakers and dependents. But as is, especially under increasing neoliberalization, the goal seems to be to cut all social supports out and let free market competition sort it.

From within this moral system, modern consumer capitalism is morally praiseworthy because it has led to increased freedom, equality, efficiency, independence, technological advances, and social progress. According to Steven Pinker: "The Enlightenment, moreover, worked: We live longer, healthier, safer, wealthier, freer, more peaceful and more stimulating lives than those who came before us. And by "we" I don't just mean we in the West. This progress is encompassing the world."[53]

From outside this moral system, things appear quite differently. Here it worth revisiting the quote from the Osage chief Big Soldier I quoted in the last chapter:

> I see and admire your manner of living, your good warm houses, your extensive fields of corn, your gardens, your cows, oxen, workhouses, wagons, and a thousand machines that I know not the use of. I see that you are able to clothe yourselves, even from weeds and grass. In short, you even do almost what you choose. You are surrounded by slaves. Everything about you is in chains, and you are slaves yourselves. I fear if I should exchange my pursuits for yours, I too should become a slave.[54]

There is no doubt that modern science and capitalism have allowed for significant improvement in health, comfort, and abundance, but Big Chief claims this is made possible only by slavery. In light of Wildcat's commentary, we could take this to mean not only the destruction of the natural environment, the death of the cosmos and the death of god but also the enslavement of humanity itself to technology, work, and consumer goods.

High confidence in the superiority of modern consumer society in light of the extensive ongoing destruction of the environment and the exploitation that drive the modern economy is only possible through what Zygmunt Bauman calls "adiaphorization." This refers to the "exemption of certain conducts and certain aspects of interrelation and interaction from ethical significance and so obliquely denying their potentially violent character."[55] In other words, adiaphorization means the erasure of certain moral obligations. The atomism of modern society presents a world of autonomous individuals who are morally responsible to themselves and erases the fundamental dependencies and connections we have to other human beings, the world (in the Arendtian sense) and the Earth. As Bauman puts it:

> The concepts of responsibility and responsible choice, which used to reside in the semantic field of ethical duty and moral concern for the Other, have moved or have been shifted to the realm of self-fulfillment and calculation of risks. In the process, the Other as the trigger, the target, and the yardstick for a responsibility accepted, assumed and fulfilled has all but disappeared from view, having been elbowed out or overshadowed by the actor's own self. "Responsibility" means now, first and last, *responsibility to oneself* ("You owe this to yourself," as the outspoken traders in relief from responsibility indefatigably repeat), while "responsible choices" are, first and last, such moves as serve well the interests and satisfy the desires of the actor and stave off the need to compromise.

> The possibility of populating the world with more caring people and inducing people to care more does not figure in the panoramas painted in the consumerist utopia. The privatized utopias of the cowboys and cowgirls of the consumerist era show instead vastly expanded 'free space' (free for *my* self, of course)—a kind if empty space of which the liquid-modern [post-modern] consumer, bent on solo performances and solo performances only, never has enough. The space that liquid-modern consumers need and are advised from all sides to fight for can be conquered only by evicting other humans—and particularly the kind of humans who care for others or may need care themselves.
>
> It is just as Emmanuel Levinas adumbrated, when musing that rather than being a contraption making peaceful and friendly human togetherness achievable for inborn egoists (as Hobbes suggested), society may be a stratagem for making a self-centered, self-referential, egoistic life attainable for inborn moral beings, through cutting down the responsibilities for others that go together with the presence of the Face of the Other, indeed, with human togetherness.[56]

Here Bauman claims that modern individualism requires the elision of our interconnectedness and mutual responsibilities. To think that we owe nothing to nobody, and we are responsible primarily for and to ourselves requires the positing of a self-contained subject and the silencing of the traditional moral imperatives to care for and respond to each other. Bauman claims that unlike Hobbes' account of egoistic people being organized to bring about peace, what is more likely is that modern society is taking children of mothers and teaching them to be instrumentally rational egoists. In other words, *Homo economicus* is not born but made.

We don't take general education courses on "How to be an Individual." We don't need to. The moral assumptions and ideals that silence our connectedness and stress self-reliance and individual freedom are "in the water"—they have become natural and taken for granted. This is remarkable considering how atomistic individualism is so radically different from traditional moralities and understandings of human nature. We could say that this atomism functions like an epistemology of ignorance.

Here I am appealing to the work of Charles W. Mills and Linda Alcoff who have argued that ignorance (in this case, ignorance of our dependency, interconnectedness, and moral responsibilities) is often taught. In other words, ignorance is not just a lack of knowledge, but sometimes a learned lack of knowledge or a selective understanding that purposefully elides important details or facts. Mills argues that with regard to race we find "structural group-based miscognition"—ignorance about race issues that have become natural.[57] In other words, members of a particular group are sometimes not taught certain information and funneled into certain perspectives that misrepresent reality. Specifically Mills has explored what he calls "white ignorance"—the ways that selective history, racist stereotypes, and so on can come together to form a deeply ignorant understanding of race, race relations, racist history, and racial identity. This is the sort of thing we see at work when some claim the American Civil War was not about slavery or that monuments to the South are about tradition and not racism. These views are really bizarre to an outsider, but the same thing can be said about modern individualism.

The same thing happens with the atomistic understanding of society. No society is this way and yet it is widely accepted as the nature of society and the individual. No society *could* be this way, and yet it is taken as the right way to be. Kittay argues that one of the results of the unquestioned ubiquity of these ideals is that many feminists have sought to help women have more independence, when Kittay thinks we should be challenging the ideal of autonomy and independence and helping men recognize that they need to be engaging in more dependency and care work.[58] Arlie Russell Hochschild makes a similar claim:

> When in the mid-nineteenth century men were drawn into market life and women remained outside of it, female homemakers formed a moral brake on capitalism. Now American women are its latest recruits, offered membership in the public side of market society on the same harsh terms as those offered to American men. The result makes for a harshness of life that seems so normal to us we don't see it. We really need, I believe, a revolution in our society and in our thinking, one that rewards care as much as market success, one that strengthens a nonmarket public sphere—like the old village commons.[59]

Neither Kittay nor Hochschild are calling for a return to traditional gender roles. They are calling for a rejection of atomism, a reconception of the self, and the radical reorganization of society in light of the reality of dependency and the necessity of care. Both agree that modern society is crudely harsh toward these things and are attempting to shatter the veneer of normalcy of atomism.

Despite the barrage of powerful critiques from many sides aimed at the many heads of the modern hydra (atomism, consumerism, modern epistemology, scientism, positivism, and so on), the hold of atomism is only getting stronger. The last forty years have seen the rise of neoliberalism which has accelerated the trend toward hyper-individualism, entrenched the modern moral assumptions and ideals of independence, freedom, autonomy, and self-reliance and globalized consumer capitalism. And with this entrenchment comes a worrisome pervasive cultural thoughtlessness about these issues.[60]

Giddens describes modernity as a juggernaut that cannot be stopped.[61] This may be the case. The abundance of consumer choice, the allure of technological advancement, and the legitimate progresses of modernity obscure the ongoing exploitation of workers, our world alienation and the death of cosmos. But there is one thing that cannot be elided and that could eventually stop modernity whether we want it to or not: the destruction of our Earth.

DESTROYING OUR NEST

Bonnie Mann takes the arguments that Kittay and Hochschild make about dependency and care and applies them to the Earth.[62] Atomistic independence, autonomy, and self-reliance are all an illusion. All humans are born to a mother. All humans learn a shared language. In highly interconnected advanced capitalism, there are many people each of us depend on to produce the many things we use daily—most obviously food. Of course, on a broader level, neither mothers, nor language, nor communities, nor food would exist without the Earth.

Environmentalists and environmental thinkers have long warned about the recklessness of the modern juggernaut. There are many different environmentalisms, and they are not all in agreement; but in their varying concerns about the Earth, the environment, animals, plants, and so forth, they are worrying and caring for something that goes beyond humans. As such environmental thought tends to be uniquely sensitive to the anthropocentrism of modernity. Even then, there are still many environmentalists who frame their concern for the environment in terms of human needs ("We are destroying *our* own nest!"). Some strands of environmentalism like Deep Ecology and Ecocentrism reject what they call "shallow ecology" and claim that unless we reject anthropocentrism and the ensuing privileging of human needs over animals, plants, the oceans, the air, and so forth, then it is unlikely we will recognize the depth of the environmental crisis.[63] The distinction between "deep" and "shallow" is meant to draw attention to the ways that too many environmental approaches settle for small gains—building parks, recycling, raising awareness, and so forth—but do not radically challenge consumer capitalism, modern atomism, and modern anthropocentrism.

Sessions, for example, claims that if we were to take environmental destruction seriously and truly seek to reverse it, we would have to radically restructure human society, moving most humans to large cities and leaving one-third of the land to wilderness (and rewilding) and another one-third to land with some farming mixed with unfarmed land.[64] This sort of proposal is going to be dismissed as unpractical and absurd, but even if it is, it shows not only the scope of the environmental problem but also the hold that our anthropocentrism has on us.

While I will not systematically address environmental literature, I hope it is clear from our discussion both in the first chapters on love of the cosmos and in the last on love of the Earth and Nature (or all the people of a particular place) that for many premodern and nonmodern peoples humanity was intimately connected to and porous with the natural world around us. Robin Wall Kimmerer offers a trenchant analysis:

> The sanitized suburban life has succeeded in separating us from the plants that sustain us. Their roles are camouflaged under layers of marketing and technology. You can't hear the rustle of corn leaves in a box of Foot Loops. Most people have lost the ability to read the role of a medicine plant from the landscape and read instead the "directions for use" on a temper-proof bottle of Echinacea. Who would recognize those purple blossoms in this disguise? We don't even know their names anymore. The average person knows less than a dozen plants, and this included such categories as "Christmas Trees." Losing their names is a step in losing respect. Knowing their names is the first step in regaining our connection.[65]

Robin Wall Kimmerer's book, *Gathering Moss: A Natural and Cultural History of Mosses*, explores the small and forgotten world of mosses. Here moss serves as a metaphor for everything in the natural world that is connected to human existence and yet which we ignore and even exploit. In the crushing final chapters, she describes the moss trade—yes, even moss, which may take hundreds of years to develop, is stripped from trees in the Pacific Northwest and sold all over the world.

Ecology has shown the ways that entire ecosystems are connected, but in our ignorance, lack of interest, anthropocentrism, and greed, modern humans radically and repeatedly disrupt and destroy these sometimes delicate interconnected systems to make a profit. I will not post here the disturbing statistics about species die-off, destruction of the rain forests, global warming, plastic in the oceans, coral bleaching, and so forth. It is widely known that this is happening. And yet, to summarize Donald Trump: economy first.[66] This means, of course, economy before humans and humans before the environment.

I'm not sure if there will ever be enough facts to alter the modern formula of economy > humans > environment. Marx and Critical Theorists have shown the modern subsuming of humans to capital. Environmentalists have shown the modern subsuming of the environment to humans and capital. The facts of the exploitation of labor, nature, and the Earth are everywhere, and yet the comforts of modern life and the hermetically sealed hyper-individualism of modernity isolate us in ignorance and indifference. I am not suggesting environmentalists should stop shouting their criticisms from the rooftops. They should be louder. I also recognize that many environmentalists are attempting to lead by example, seeking to live life more sustainably and with greater love for the environment. In the writing of diverse thinkers such as Thoreau, Robin Wall Kimmerer, David Abrams, John Muir, Val Plumwood, and so on, there is a palpable love of the world. It seems to me that *amor mundi* is the glue that ties the critical message in with the positive one.

NOTES

1. Jonas, Hans. *The Phenomenon of Life: Toward a Philosophical Biology* (Evanston, IL: Northwestern University Press, 2001).

2. Brague, Rémi. *The Wisdom of the World: The Experience of the Universe in Western Thought* (Chicago: University of Chicago Press, 2003).

3. Heidegger, Martin. *The Question Concerning Technology and Other Essays* (New York: Harper Perennial, 1977).

4. Tucker, ed., *The Marx and Engels Reader (Second Edition)* (New York: W.W. Norton & Company, 1978).

5. Riesman, David. *The Lonely Crowd: A Study of the Changing American Character* (New Haven: Yale University Press, 2001). Sociology in particular has been very sensitive to increasing individualism in modernity.

6. Singer, Irving. *The Nature of Love, Vol. 1: Plato to Luther* (Chicago: University of Chicago Press, 1987).

7. See Ch 1.

8. Singer, Irving. *The Nature of Love: Courtly and Romantic* (Boston, The MIT Press, 2009).

9. Singer, Irving. *The Nature of Love: The Modern World* (Boston, The MIT Press, 2009), 10.

10. Singer, *Nature of Love: The Modern World*, 10.

11. Seidman, Steven. *Romantic Longings: Love in America, 1830–1980* (New York: Routledge, 1993).

12. Ewen, Stewart. *Captains of Consciousness: Advertising and the Social Roots of the Consumer Culture* (New York: McGraw-Hill, 1976).

13. Strasser, Susan. *Satisfaction Guaranteed: The Making of the American Mass Market* (Washington, DC: Smithsonian Institution Press, 1989). See also footnote 12.

14. Taylor, Charles. *A Secular Age* (Cambridge, MA: Harvard University Press, 2007).

15. Seidman, *Romantic Longings*; Coontz, Stephanie. *Marriage, a History: How Love Conquered Marriage* (New York: Penguin, 2005).

16. For more details, see the three volume study of love in Western philosophy by Irving Singler.

17. Soble, Alan, ed. *Sex, Love and Friendship* (Amsterdam: Editions Rodopi, 1997).

18. Leigh, McEvoy, Adrianne, ed., *Sex, Love and Friendship: Studies of the Society for the Philosophy of Sex and Love: 1993–2003* (Amsterdam: Rodopi, 1997).

19. https://www.theatlantic.com/magazine/archive/2018/12/the-sex-recession/573949/?utm_source=facebook&utm_campaign=the-atlantic-fb-test-584-1-&utm_content=edit-promo&utm_medium=social&fbclid=IwAR0G-NNxMgshh2Zwy2csblwLPm3zZJjIkIh_CBwyLBI9GG4qSiEEOVSZ_c8. Accessed Jan 1, 2019.

20. Tönnies, Ferdinand. *Communities and Society* (New York: Dover Publishing, 2011); Durheim, Emile. *The Division of Labor in Society* (New York: Free Press, 2014).

21. Riesman, *The Lonely Crowd.*

22. Elias, Norbert. *The Society of Individuals* (London: Continuum, 2001).

23. Putnam, Robert. *Bowling Alone: The Collapse and Revival of American Community* (New York: Touchstone Publishing, 2001); Bauman, Zygmunt. *The Individualized Society* (Malden, MA: Blackwell Publishing, 2001).

24. Taylor, Charles. *Philosophical Arguments* (Cambridge, Harvard University Press, 1997), vii.

25. Nietzsche, Friedrich. *The Genealogy of Morals* (New York, NY: Oxford University Press, 1999).

26. Lakoff, George and Mark Johnson. *Metaphors We Live By* (Chicago: The University of Chicago Press, 1980).

27. Taylor, *Philosophical Arguments*, 7.

28. Taylor, Charles. *Philosophical Papers: Volume 2, Philosophy and the Human Sciences* (New York, NY: Cambridge University Press, 1985).

29. Taylor, Charles. *The Language Animal: The Full Shape of the Human Linguistic Capacity* (Cambridge: The Belknap Press, 2016).

30. Taylor, *The Language Animal.*

31. While there are rogue branches of economics that recognize humans are more complex than mainstream economics, they are not widely influential.

32. Castro Gomez, Santiago. "The Social Sciences, Epistemic Violence and the Problem of the 'Invention of the Other'." *Views from South*, Vol. 3, No. 2, 2002, pp. 269–285. and Foucault, Michel. *Discipline and Punish* (New York: NY: Vintage Books, 1995) and Foucault, Michel. *Power/Knowledge* (New York: Vintage, 1980); Scott, James C. *Seeing Like A State: How Certain Schemes to Improve the Human Condition Have Failed* (New Haven, CN: Yale University Press, 1998).

33. Taylor, Charles. *The Ethics of Authenticity* (Cambridge: Harvard University Press, 1991).

34. Taylor, *The Ethics of Authenticity* , 2–3.

35. Horheimer, Max and Theodor Adorno. *Dialectic of Enlightenment: Philosophical Fragments* (Stanford: Stanford University Press, 2002).

36. De Toqueville, Alexis. *Democracy in America* (New York: Penguin Classics, 2003).

37. Lloyd, Genevieve. "The Man of Reason." *Metaphilosophy*, Vol. 10, No. 1 (January 1979). Mills, Charles W. *The Racial Contract* (Ithaca: Cornell University Press, 1997).

38. There are many works on this topic. See James C. Scott, Charles Mills and Vine Deloria Jr.

39. There are also many works on this topic. See, for example, *Ehrenriech, Barbara. For Her Own Good: Two Centuries of Experts Advice to Women* (New York: Anchor, 2005).

40. Bonilla-Silva, Eduardo. *Racism without Racists: Color-blind Racism and the Persistence of Racial Inequality in America* (Lanham, MD: Rowman and Littlefield, 2017).

41. Fraser, Nancy and Linda Gordon, "A Genealogy of Dependency: Tracing a Keyword of the U.S. Welfare State", in Eva Feder Kittay and Ellen K. Feder, eds. *The Subject of Care: Feminist Perspectives on Dependency* (Lanham, MD: Rowman and Littlefield, 2002).

42. Arendt, Hannah. *The Human Condition* (Chicago: University of Chicago Press, 1998).

43. Fraser and Gordon, "A Genealogy of Dependency: Tracing a Keyword of the U.S. Welfare State", 16.

44. Ibid.

45. Mann, Bonnie. "Dependency on Place: Dependency in Place", in Eva Feder Kittay and Ellen K. Feder, eds., *The Subject of Care: Feminist Perspectives on Dependency* (Lanham, MD: Rowman and Littlefield, 2002).

46. Kittay, Eva Feder. *Love's Labor: Essays on Women, Equality and Dependency* (New York: Routledge, 1999), 23.

47. Kittay, *Love's Labor.*

48. Ibid., 14.

49. Smith, Dorothy. *The Everyday World as Problematic: A Feminist Sociology* (Boston: Northeastern University Press, 1989).

50. Kittay, *Love's Labor*, 15

51. Fraser and Gordon, "A Genealogy of Dependency: Tracing a Keyword of the U.S. Welfare State."

52. Taylor, *Philosophical Papers.*

53. https://www.nytimes.com/2018/04/10/opinion/steven-pinker-enlightenment-now.html?module=inline. Accessed Jan 1, 2019.

54. Tinker, George. "Religion", in Frederick Hoxie, ed., *Encyclopedia of North American Indians* (New York: Houghton, 1996).

55. https://lareviewofbooks.org/article/disconnecting-acts-interview-zygmunt-bauman-part/#!. Accessed Jan 1, 2019.

56. Bauman, Zygmunt. *Does Ethics Have a Chance in a World of Consumers?* (Cambridge: Harvard University Press, 2008), 52–53.

57. Sullivan, Shannon and Nancy Tuana, eds. *Race and Epistemologies of Ignorance* (Albany: SUNY University Press, 2007). 13.

58. Sullivan and Tuana, eds. *Race and Epistemologies of Ignorance.*

59. Hochschild, Arlie Russel. *The Commercialization of Intimate Life: Notes from Work and Home* (Berkeley: The University of California Press, 2003), 8.

60. I have examined the relationship of neoliberalization and thoughtlessness in a previous work. Pack, Justin. *How the Neoliberalization of Academia Leads to thoughtlessness: Arendt and the Modern University* (New York: Lexington Books, 2018).

61. Giddens, Anthony. *The Consequences of Modernity* (Stanford: Stanford University Press, 1991).

62. Mann, "Dependency on Place: Dependency in Place."

63. See Naess, Arnie. "The Shallow and the Deep, Long-range Ecology Movement: A Summary." *Inquiry*, Vol. 16, No. 1, 1973, 95–100; Devall, Bill and George Sessions. *Deep Ecology: Living as if Nature Mattered* (Layton, Utah: Gibbs Smith, 2001).

64. Sessions, George, ed. *Deep Ecology for the Twenty-First Century* (Shambhala, 1995).

65. Wall Kimmerer, Robin. *Gathering Moss: A Natural and Cultural History of Mosses* (Corvallis, OR: Oregon State University Press, 2003), 101–102.

66. Trump's motto has been "America First," but when it comes to exploitation and the environmental crisis he has made it clear that the economy comes before all. This is a very American sentiment.

Conclusions

Love the World!

Learning to love the world is important on multiple levels. It is important to the Earth itself and to the plants, animals, water, and air here on the Earth. But it is also important to human happiness on an existential level. *Amor mundi* is a task that should be relevant not only in our most selfish concern for our own happiness but also in our most compassionate concerns for other humans and the Earth. Let me summarize the "payoffs" of learning to love the world before making some brief suggestions about inculcating *amor mundi*. First, I will turn to Arendt's claims about public happiness and then to Deloria and Wildcat's claims about the environment.

1. According to Arendt, modern consumer society (America in particular in *On Revolution*) has chosen to aim for private happiness instead of public happiness. As we saw in chapter 4, the lost treasure that was the goal of the ancient polis and that occasionally resurfaces in revolutionary moments is public happiness—moments of deep commitment and purpose with others.

While it is true that modern consumer society does enable the opportunity for increased private happiness, it does so at the expense of public happiness. The institutions, spaces, and relationships that are needed for public happiness are not compatible with the abundance-maximizing machinery of modern consumerism. As a result, for Arendt, we have eaten away (consumed) our common, public worlds and have been left with abundance but alienated and isolated. Importantly for Arendt, the hyper efficiency of the modern economy invades all spaces and creates a perpetual competitive struggle that results in widespread bullshitting (BS) and strategic maneuvering. In rare revolutionary moments, this competitive BS dissipates and a more genuine, public spirit emerges.

For Arendt, freedom is not found in the workplace. Freedom is found in spaces where no one is forced to do anything. On this definition of freedom, the modern world is actually less free in many ways than certain peoples in the premodern world. The Greek polis, on Arendt's reading, was a purposefully created BS-free space where citizens could come to participate in a common world. Such a place gives us meaning, purpose, recognition, and community—that is to say—public happiness.

Critics of Arendt may claim that the polis was not such a place, that strategy exists everywhere, and so forth. But they may miss the philosophical importance of her claim: we have documented moments of such freedom (revolutionary moments), and we can create such places of freedom and public happiness if we are committed to it. If Arendt is correct, then learning to love and care for a common (human) world is both a way to obtain public happiness and to be free. Private economic happiness, on the other hand, seems to come at great expense and requires socially eliding the exploitation and alienation of the modern economy: the structuring of society to maximize private happiness has produced an oddly inhuman society. Furthermore, democracy itself is undermined by this commitment to private happiness. A healthy democracy, freedom, and public happiness are all related on Arendt's account, and we cannot have any of them in a divided consumerist and atomistic society. Existentially, if we want to be happy, we must do it with others.

2. While love of a (human) world helps us to overcome our isolation with other humans, love of the Earth, or the particular peoples (plants, animals, etc.) of a particular place can help humans reconnect with their natural environments and develop deep and loving relationships with them. This is also a task of love, which requires recognizing our dependence on the Earth, radically changing our current consumerist lifestyles, and actively seeking to reconnect with place. There is joy in connecting with place in a way that is existentially relevant, but there is also the nonanthropocentric goal of respecting life and preserving it.

Amor mundi in all three forms we have examined is about recognizing our interconnection and dependency with the world (cosmos, Earth, and human worlds). It is an ethical responsibility to love and care for the world. But even if these claims are accepted, our modern atomism is deeply entrenched in our understanding, including our understanding of freedom, the self, society, and morality. It is not simply a matter of sluffing these modern understandings off when we want and stepping into a nonmodern or postmodern alternative. There is no easy recipe for overcoming the hold of the modern hydra and far better guides for learning to love the world than myself (such as Deloria and Wildcat, Robin Wall Kimmerer and Arendt). My goal in the book has been to articulate what is meant by *amor mundi* and show the need for it. This still leaves the details of the practical task of learning to love the world rather open-ended. While a complete treatment of the diverse and fascinating discussions of overcoming modern atomism and learning to love the world would take another book, I do want to finish by recognizing some traditions and practices that could help recover or develop a love of the world.

RECOGNIZING ALTERNATIVE TRADITIONS

Stated bluntly, if modernity has undermined *amor mundi*, then attempts to think and recover some kind of *amor mundi* will likely involve moving beyond or looking outside of modernity. This text has already centered on three pre/nonmodern kinds of love of the world (the Timaean cosmos, the ancient Greek polis and Native American place-world), but there are of course many others traditions that I have not specifically addressed. While not all premodern cultures and traditions have a form of *amor mundi* (the Gnostics seem to manifest a hatred of the world), even a cursory familiarity with anthropology shows that many do.

By treating itself as the peak of human understanding and knowledge, modernity dismissed nonmodern traditions. Speaking not only of Kant but also atomistic modernity in general, Lorraine Code claims:

> In placing man at the center of the universe it tacitly promoted a picture of the world, both physical and human, that privileged and was subservient to a small class and race of people whose sex required no mention because it was presumptively male and in any case irrelevant and who were uniformly capable of achieving a narrowly conceived standard of rationality, citizenship, and morality.[1]

Modernity enthroned the assumptions and ideals of a very narrow group of people, who then sought to remake the world in their own image. This included descriptions of what it is to be human (autonomous [male] individual), what constitutes rationality (calculative or instrumental rationality), how humans should relate to each other, and so forth. To be clear, modernity is not a monolithic tradition, and there have always been ongoing disputes and challenges throughout modern intellectual history.[2] Nonetheless, there has also been a common and consistent tendency in the modern tradition to reject and dismiss nonmodern and premodern traditions as primitive and unimportant except as historical phenomena.

The project of modernity did not just dismiss other traditions but has radically changed human existence in ways that affect not only the small class and race of people who articulated and implemented it, but everyone across the entire globe—whether they wanted to be changed or not. Thus, there is a double injustice: the rejection and silencing of other traditions, followed by the imposition of modernity onto everyone. The late twentieth century has seen the rise of a series of "new social movements" like feminism, critical race theory, and postcolonialism that criticize the way women, nonwhite and non-Western peoples and traditions were/are treated as inferior, excluded, ignored, and altered by atomistic modernity.[3] These traditions often have in common the criticism that the world being described and proscribed by modernity is neither a good fit with the lived experience of the individuals in these traditions (it is a bad description) nor as morally compelling as modernity seems to think (a bad prescription). As such, these silenced traditions often want to be allowed to speak on their own terms, in their own voice, and to decolonialize and recover their own practices.

These nonmodern traditions, like modernity itself, are not monolithic and have diverse responses to modernity. From the perspective of modern world alienation, most, if not all of them, have important and interesting things to teach us. I have examined one such tradition, Native American metaphysics, and ethics, but the analysis of *amor mundi* and world alienation would only grow richer if we examined African-American philosophies, African philosophies, Asian philosophies, Latin American philosophies, Queer studies, and so forth. To do so goes beyond the space that I have here, but I want to all-too-briefly recognize and point to how these traditions can help us think about recovering *amor mundi*.

The environmentalist Bill Devall claims that as environmental thinkers sought alternatives to the "dominant paradigm" of modernity they turned to these pre- and nonmodern traditions.[4] He claims there were five sources of the Deep Ecological environmental approach he espouses: (1) Eastern traditions, (2) Native American Traditions, (3) minority and self-critical traditions within Western thought itself, (4) the scientific discipline of ecology, and (5) art and artists.[5] (He was correctly taken to task for not recognizing feminism and eco-feminism on his list.)[6] It is not surprising that disillusioned Western thinkers would turn away from modernity toward the traditions that it had excluded, but there are reasons to be careful with this turn: recall Vine Deloria Jr.'s description of spiritually lost white folks bumming around Native Americans looking for spiritual advice and his frustration with whites seeking Native American solutions to white problems. His point was not that Native Americans might not have good advice and insight into world alienation and the environmental crisis, but that desperate whites often misunderstood and misrepresented Native American perspectives and too often failed to address the Western cultural sources of these problems.

While recognizing these difficulties, I want to nonetheless briefly discuss some of the particularly salient things we can learn about *amor mundi* and overcoming world alienation from some of these traditions. I cannot offer an exhaustive overview, but my hope is that the reader that is not familiar with these issues will search out these sources and read more for themselves.

1. Feminism in particular stands out as a tradition that has offered insightful critiques of modernity. Already in the last chapter, we have seen how feminist thinkers have criticized the way modernity continues to propagate the problematic dichotomy of rationality and emotion that we find in Plato and the Stoics. Counter to this, feminists have shown the importance of emotion and love in knowing and ethics.[7] We saw a similar complaint from the Osage chief Big Soldier that while modernity brings power, it enslaves the world. This is a repeated concern in anthropological and postcolonial literature: the abstraction, fragmentation, and emptiness of Western culture. While not always framed in the terms I have been using, there is a common lament about the loss of love and respect for the world.
2. Additionally, feminists and postcolonial thinkers have also further developed Marxist standpoint theory, which claims that those who are damaged or excluded by a particular society are in an epistemologically privileged position to understand that society.[8] If a particular form of social organization is structurally privileging a certain group, then this group is more likely to be content with the status quo and less likely to challenge it. At worst, they may not even see the inequalities and injustices that make their own advantages possible. Marx claimed that the proletariat was in a unique position to understand the injustices of capitalism because these negatively affected the proletariat in a way that made them more understanding of the dark side of capitalism. Feminism expands this point to women and any other group that has been colonized, diminished, or marginalized in a modern society. Standpoint theory, then, offers the epistemological and moral justifications we need for returning to the traditions that were marginalized by modernity. Not surprisingly, there are many colonialized traditions that are seeking to decolonialize their own cultures and bring attention to the too-often unrecognized underside of modernity. There are rich dialogues challenging modernity that come from African-American thinkers, African thinkers, Asian thinkers, Latin American thinkers, queer thinkers, and so on.[9]
3. Importantly, many feminist and postcolonial thinkers advance relational epistemologies. Traditional scientific approaches tend to narrowly reduce their objects of study into their physical aspects by stripping away the many registers of meaning that suffuse our world in everyday action,

> but relational epistemologies refuse to do so and instead emphasize and explore the complex interrelations that make up systems and the situated location of the knower in such complex systems. This implies that knowing is not a solitary task guided by a sure method, but one that happens with others complex, multidimensional systems. A relational epistemology, therefore, pay careful attention to the situatedness of the knowers involved and the multiple registers of meaning at play. In environmentalism, the relational approach that has served as a primary inspiration is ecology, a discipline that focuses on the interrelationships between all the aspects of an ecosystem. In feminist and postcolonial thinkers, we find a similar attempt to recognize that the issues being thought and addressed by these thinkers always occur in a hypercomplex nexus of social, political, ethical, gendered, racial, economic, and aesthetic meaning. In other words, any particular issue is simultaneously environmental, social, political, ethical, feminist, racial, economic, aesthetic, and so forth. Instead of attempting to jump out of this web by focusing only one aspect of the phenomena in question, it must be responsibly approached by knowers working together in relation to all these registers insofar as is possible. On the one hand, this is impossible to do perfectly, but feminist and postcolonial thought has shown that thinking together in interdisciplinary and intersectional terms can be very rich and powerful.

Take, for example, the classic work *Borderlands/La Frontera* by Gloria Anzaldúa.[10] Anzaldúa's groundbreaking approach begins from her epistemic location: a lesbian Latina woman living, working, thinking, and exploring the tensions that occur by virtue of being *mestiza*: a member of multiple, mixed worlds. Anzaldúa was raised in a borderland both literally and metaphorically. Literally because she was raised on the US side of the Mexican border in a Latino community. Metaphorically because she is not fully Mexican (born in the United States), not fully US American (raised in a Latino community in Spanish), persecuted by both cultures for her race, gender, and sexual orientation—she is caught in the space between multiple worlds, their antagonism toward each other and their shared patriarchy and homophobia. From this liminal place of exile, she explores mestiza consciousness and experience, tracking the tensions placed upon her and moving out into alternative ways of knowing (exploring mystical experiences and using poetry to think in these outcast spaces). *Borderlands/La Frontera* is simultaneously autobiography, epistemology, ethics, social criticism, poetry, and history.

Following the lead of thinkers like Anzaldúa, relational epistemologies refuse to retreat into a method or reductive approach and instead seek to stay with these tensions. This does not mean that one has to reject science, but rather that one must not reject the tensions that come from the many registers of meaning in order to do "only science." Indeed, with new fields like political ecology, we see scientists beginning to attempt to do science with more attention to how their scientific discoveries relate to these complex fields of meaning.[11] Especially in light of our environmental crisis, science denial and political attacks, scientists are increasingly having to do more than "just science." The rise of new interdisciplinary approaches like political ecology is heartening and exciting because in some cases it represents a movement outwards by scientists from the established center toward these complex registers of meaning that have typically been the realm of the humanities. This does not mean, however, that marginalized traditions are being appreciated and recognized by modernity. As Deloria points out, nonmodern traditions are often only allowed to speak if they are supported by established white scholars.[12] Someone really exploring liminal spaces like Anzaldúa is likely to remain influential only in certain disciplines of the humanities.

There are an increasing number of scholars who work both inside modernity while being outside it. Robin Wall Kimmerer, to name one figure we have already discussed, is a Native American scientist. She is able to speak both "languages" and in her nonscientific works explores the tensions she deals with as someone who is both "in" and "out." Figures like Robin Wall Kimmerer are still the exception to the rule, but hopefully the future will have more thinkers like her and more interdisciplinary approaches like political ecology that connect science up with the registers of meaning that are normally the terrain of the humanities and social sciences. Unfortunately, at the same time these positive developments are occurring, the disciplines that have been the most receptive to marginalized traditions (especially the humanities) are in decline.

In terms of *amor mundi*, fields such as religious studies and theology, ethic studies, anthropology, art, history, philosophy, and so forth are of pivotal importance, but they are currently struggling for their very existence. At the same moment, some scientists are recognizing the need to get politically involved and address more fully the multifaceted implications of their research, the disciplines that focus on these different registers of meaning are being replaced by STEM and business programs. Frankly, under the neoliberalization of the university, it seems to be one step forward, two steps back.[13]

This discussion is too brief, but does indicate the importance of marginalized traditions (and disciplines). Overcoming modern atomism and developing *amor mundi* will likely require a sophisticated historical, comparative, interdisciplinary approach that seeks to learn from traditions that are not as world alienated as modern capitalist cultures. Furthermore, it will require changing our everyday practices. I want to finish by pointing to three examples of practices that could contribute to overcoming atomism and developing *amor mundi*. These come from contemporary philosophy, premodern tradition, and political activism.

FOCAL THINGS AND PRACTICES

Albert Borgmann is a philosopher that has given thought to recover practices that help overcome world alienation. Borgmann articulates the questions we have been dealing with in terms of technology and claims that technology tends to separate us from the world. He gives the example of food:

> For most of us, the practices of agriculture have entirely disappeared into the machinery of production. The commodities of food appear on supermarket shelves as if from nowhere, with fake reminders of the farms we know from children's books. And this pattern of technology becomes visible in its stark and general two-sidedness of commodification and mechanization. Commodification is the detachment of things and practices from their traditional contexts, and it is the conversion of things and practices into freely available commodities. Mechanization is the replacement of traditional contexts and competencies by increasingly powerful and concealed machineries. As workers we indenture ourselves for the requirements of the machinery. As consumers we revel in the abundance of unencumbered pleasures.[14]

The supermarket stands between us and the origins of our food, placing us at a remove and disconnecting us from the place where our food is grown, from the people who grew it, from the time when it grows, and so forth. Like Arendt, Deloria, and the other thinkers we have been examining, Borgmann sees this disconnection as fundamental to modernity:

> ...the culture of technology has finally depressed the glory and the misery of the human condition to distraction and indecision. For most people in this country, the overt challenges of global warming and global justice are uncertain specters in the distant background. The profound challenge of the good life is an ever postponed task. The foreground of life is occupied with worries about the stability of work and the little and quickly fading thrills of consumption. As members of the technological society, we have systematically uprooted the relations that once had grounded our lives in a certain community, a definite place, and an overarching time. The machineries that now support us fail to engage us, and the commodities that are supposed to please us have turned out to be joyless.[15]

But Borgmann argues we can actively pursue ways to reconnect with community, place and time. This can be done through what he calls "focal things and practices."[16] The word "focal" here is meant to draw attention to the way that these things and practices can bring our life into focus and help us sort through the distractions, making the world clear in light of the particular things or practice. Borgmann gives the example of running. He points out that committed and serious runners often find that running simplifies and helps organize their lives. The activity of running gives them time to think. Their feet running over the dirt, soil, and pavement connects them to the Earth and place. Many other intensive practices like biking, guitar playing, gardening, chess, skateboarding, and so forth can function in a similar fashion; they provide purpose, organization, and clarity to life. The surfer working on her craft no longer obsesses with phone apps—instead of being driven by distractions, her technology use is now defined by the focal practice of surfing.

Such activities often involve focal things: running shoes, a surf board, a guitar—these are things that are the center of these practices. These can obtain a kind of sacredness or become beloved. Focal things could be a family heirloom like a wedding ring or silverware that has been passed on for generations, a table, a vintage car, a recipe for jam, a relic in a church, a historical building. Focal practices and things are, to put it in Arendtian terminology, semipermanent parts of a world. The world is impoverished if they are destroyed and thus they should be respected and cared for.

Borgmann's point is that we can seek to create focal practices and focal things in our life and reconnect with each other, with the (Arendtian) world and with the Earth. Importantly, this is something that many of us have likely already experienced in our lives at some point or another. In other words, Borgmann argues that for all our world alienation, we remain connected to the Earth and to each other, and we can foster these connections and actively seek to make new connections. We can, as Robin Wall Kimmerer hoped, learn the name of plants. We can learn the history of our place, eat seasonally, get involved in the community, and so forth. There are many different ways to reconnect to each other, to the physical and cultural (Arendtian) world, to the environment—in short, to place.

GIFTS

The kind of connected focal relationships that Borgmann (and Robin Wall Kimmerer, Deloria and Wildact, and Brague) is describing between ourselves and the world promote responsibility, care, and love. They have the power to produce in us deep feelings of gratitude.

The poet Lewis Hyde has described this relationship as a gift relationship.[17] Anthropologists have long claimed that many premodern societies function as "gift economies" and the image of premodern gift economies has been leveraged as a romantic foil to the modern exchange economy. These efforts have often been problematic and at worst colonial, too often failing to understand how gift economies actually function/ed.[18] Despite these issues, there is a clear difference between gift and exchange that I suspect everyone has experienced and is familiar with: this is the difference between a gift that is about relationship and connection and an exchange that is about profit. I give a gift to show I care. I exchange to get something I want. A really good gift can bring someone to tears because it reflects deep understanding, love, and relationship. A good exchange can, and often does, occur between two strangers that will never meet again.

Modern capitalism is based on exchange. While the modern economy has proven to be hyper-efficient, it turns us into calculative, competitive, disconnected strangers. Many traditional societies, on the other hand, emphasize personal and community relationships, often to the potential detriment of economic productivity. Stated bluntly: they choose community over economy.

Hyde's discussion of gifts is helpful in understanding what it would take to learn to love the world: it means switching from a calculative, exchange-oriented approach to the world ("How do I get the most of all this?") to and caring, relational approach. Hyde engages in anthropological, historical, and phenomenological accounts of gift cultures and relationships, including a fascinating analysis of how folktales reinforced premodern morals (if the faeries give you a gift, use it, share it, but don't horde it, and don't calculate how to maximize it—those who do meet a terrible fate).

As a poet, Hyde describes how poetry is a mysterious and wonderful endeavor that requires establishing a loving, reciprocal gift relationship with the world. He claims that poetry cannot be forced and that poems come at their own pace, often appearing suddenly, like a gift from nowhere. On his end, he has to make himself receptive to the poem, to put himself in a place of receptivity, openness, caring, interest, and mindfulness where it can come to him. There is no guarantee that the poetry will happen, the best Hyde can do is relate to the world properly.

I won't go into extensive details of Hyde's account, but rather encourage the readers to examine it themselves. I do think it is clear that this gift orientation is common in modern figures that love the world (Thoreau, Emerson, Muir, and so forth) and helps explain the shock and moral indignation that premodern and nonmodern people commonly express when they came/come into contact with modern capitalist societies. For those seeking to love the world, a good place to start is to foster a gift orientation toward the world and emulate figures like Robin Wall Kimmerer who are also attempting to do so.

THE POLITICS OF NATURE

I want to finish with one final and simple point. This is that if we are to overcome modern world alienation and modern atomism, we must give the world (the Earth, plants, animals, air, rivers, water, etc.) a voice in our politics. Recently in Oregon, a judge threw out a lawsuit filed by a horse against its former owner for animal abuse. The court made the following statement:

> The court grants with prejudice defendant's motion to dismiss based on a lack of standing for Justice the horse. The court finds that a non-human animal such as Justice lacks the legal status or qualifications necessary for the assertion of legal rights and duties in a court of law. . . .
>
> There are profound implications of a judicial finding that a horse, or any non-human animal for that matter, is a legal entity that has the legal right to assert a claim in a court of law. Such a finding would likely lead to a flood of lawsuits whereby non-human animals could assert claims we now reserve just for humans and human creations such as business and other entities.[19]

The judge is absolutely correct that allowing nonhumans legal rights would lead to a flood of lawsuits. This is because we exploit animals and the Earth to a deeply unethical degree. Relegating all nonhuman life to a second tier is necessary to keep our economic engines accelerating.

Bolivia, a majority indigenous country, on the other hand, has passed "The Law of Mother Earth" which gives nature equal rights to humans.[20] A similar law has been recently passed for Lake Erie.[21]

The facile dismissal of Justice the horse is par for the course. It shows how far we are from *amor mundi*. For the sake of the Earth, the many peoples on her, democracy, and public happiness, we must change our ways and learn to love the world.

NOTES

1. Code, Lorraine. *Ecological Thinking: The Politics of Epistemic Location* (New York: Oxford University Press, 2006), 3.
2. Taylor, Charles. *A Secular Age* (Cambridge, MA: Harvard University Press, 2007).
3. Code, *Ecological Thinking*, 3.
4. Devall, Bill. "The Deep Ecology Movement." *Natural Resources Journal*, Vol. 20, No. 2, 1980.
5. Devall, "The Deep Ecology Movement."
6. Salleh, Ariel. "Deeper the Deep Ecology: The Eco-Feminist Connection." *Environmental Ethics*, Vol. 6, No. 4, 1984.

7. Jagger, Alison M. "Love and Knowledge: Emotion in Feminist Epistemology." *Inquiry*, Vol. 32, No. 2, 1989, 151–176; Nussbaum, Martha. *Love's Knowledge: Essays on Philosophy and Literature* (New York: Oxford University Press, 1992); Kittay, Eva Feder. *Love's Labor: Essays on Women, Equality and Dependency* (New York: Routledge, 1999).

8. Smith, Dorothy. *The Everyday World as Problematic: A Feminist Sociology* (Boston: Northeastern University Press, 1989); Harding, Sandra, ed., *The Feminist Standpoint Theory Reader* (New York: Routledge, 2004).

9. I will not attempt to enumerate all of these traditions and they key texts of each. This information has become increasing available through readers, compilation and available online course syllabi.

10. Anzaldúa, Gloria. *Borderlands/La Frontera* (San Francisco, CA: Aunt Lute Books, 2007).

11. Perreault, Tom, Gavin Bridge and James McCarthy, eds. *The Routledge Handbook of Political Ecology* (New York: Routledge, 2015).

12. Deloria, Jr., Vine and Daniel Wildcat. *Power and Place: Indian Education in America* (Golden, CO: Fulcrum, 2001).

13. Pack, Justin. *How the Neoliberalization of Academia Leads to Thoughtlessness: Arendt and the Modern University* (New York: Lexington Books, 2018).

14. Borgmann, Albert. "Contemplation in a Technological Era: Learning from Thomas Merton." *Perspectives on Science and Christian Faith*, Vol. 64, No. 1, March 2012, 5.

15. Borgmann, "Contemplation in a Technological Era: Learning from Thomas Merton," 9.

16. Borgmann, Albert. *Technology and the Character of Contemporary Life* (Chicago: University of Chicago, 1984), 198.

17. Hyde, Lewis. *The Gift: Imagination and the Erotic Life of Property* (New York, NY: Vintage, 1983).

18. Hénaff, Marcel. *The Price of Truth: Gift, Money and Philosophy* (Stanford: Stanford University Press, 2010).

19. https://www.yahoo.com/gma/judge-throws-lawsuit-filed-horse-against-former-owner-110306703--abc-news-topstories.html?fbclid=IwAR1fviYr7It8zgQOx58RxzqBQKhuuKUo5dI06o6aOHsRHvzVAB5PGD_gEyc. Accessed Jan 1, 2019.

20. https://www.theguardian.com/environment/2011/apr/10/bolivia-enshrines-natural-worlds-rights. Accessed Jan 1, 2019.

21. https://www.smithsonianmag.com/smart-news/toledo-ohio-just-granted-lake-erie-same-legal-rights-people-180971603/. Accessed May 1, 2019.

References Cited

Anzaldúa, Gloria. *Borderlands/La Frontera* (San Francisco, CA: Aunt Lute Books, 2007).

Arendt, Hannah "Culture and Politics." *Reflections on Literature and Culture* (Stanford, CA: Stanford University Press, 2007).

———. *Eichmann in Jerusalem* (New York: Penguin Classics, 2006).

———. *On Revolution* (New York: Penguin Books, 1990).

———. "Philosophy and Politics." *Social Research*, Vol. 57, No. 1, Spring 1990.

———. *The Human Condition* (Chicago: University of Chicago Press, 1998).

———. *The Life of the Mind* (New York, NY: Harcourt Brace Jovanovich, 1978).

Atleo, E. Richard. *Tsawalk: A Nuu-chah-nulth Worldview* (Vancouver: UBC Press, 2004).

Barad, Karen. *Meeting the Universe Halfway: Quantum Physics and the Entanglement of Matter and Meaning* (Durham, NC: Duke University Press, 2007).

Bauman, Zygmunt. *Does Ethics Stand a Chance in a World of Consumers?* (Cambridge: Harvard University Press, 2008).

———. *The Individualized Society* (Malden, MA: Polity, 2001).

Bellah, Robert. "Social Science as Practical Reason," in Daniel Callahan and Bruce Jennings, eds., *Ethics, the Social Sciences and Policy Analysis* (New York: Plenum Press, 1983).

Benhabib, Seyla. "The Pariah and Her Shadow: Hannah Arendt's Biography of Rahel Varnhagen." *Political Theory*, Vol. 23, No. 1 (Feb. 1995).

Bernauer, J.W. *Amor Mundi: Exploration in the Faith and Thought of Hannah Arendt* (Springer, 1987).

Bernstein, Richard J. *Beyond, Objectivism and Relativism: Science, Hermeneutics and Praxis* (Philadelphia, PA: University of Pennsylvania Press, 1983).

———. *The Restructuring of Social and Political Theory* (Philadelphia, PA: University of Pennsylvania Press, 1978).

Bonilla-Silva, Eduardo. *Racism without Racists: Color-blind Racism and the Persistence of Racial Inequality in America* (Lanham, MD: Rowman and Littlefield, 2017).

Bordo, Susan. *The Flight to Objectivity: Essays On Cartesianism and Culture* (Albany: SUNY Press, 1987).
Borgmann, Albert. "Contemplation in a Technological Era: Learning from Thomas Merton." *Perspectives on Science and Christian Faith*, Vol. 64, No. 1, March 2012.
———. *Technology and the Character of Contemporary Life* (Chicago: University of Chicago, 1984), 198.
Bourdieu, Pierre. *Practical Reason: On the Theory of Action* (Stanford: Stanford University Press, 1998).
Brague, Rémi. *The Wisdom of the World: The Experience of the Universe in Western Thought* (Chicago: University of Chicago Press, 2003).
Brandt, Allan M. "Racism and Research: The Case of the Tuskegee Syphillis Study." *Hastings Center Report*, Report 8, No. 6, Dec. 1978, pp. 21–29.
Castro Gomez, Santiago. "The Social Sciences, Epistemic Violence and the Problem of the 'Invention of the Other'." *Views from South*, Vol. 3, No. 2, 2002, pp. 269–285.
Code, Lorraine. *Ecological Thinking: The Politics of Epistemic Location* (New York: Oxford University Press, 2006).
Coontz, Stephanie. *Marriage, a History: How Love Conquered Marriage* (New York: Penguin, 2005).
Davis, Bret W., Brian Schroeer and Jason Wirth, eds. *Japanese and Continental Philosophy: Conversations with the Kyoto School* (Bloomington, IN: Indiana University Press, 2011).
Deloria, Jr., Vine. "Kinship With the World." *Journal of Current Social Issues*, Vol. 15, No. 3, Fall 1978, pp. 19–21.
———. *Spirit and Reason: The Vine Deloria, Jr., Reader* (Golden, CO: Fulcrum, 1999).
———. *The Metaphysic of Modern Existence* (New York: Harper, 1979).
Deloria, Jr., Vine and Daniel Wildcat. *Power and Place: Indian Education in America* (Golden, CO: Fulcrum, 2001).
De Toqueville, Alexis. *Democracy In America* (New York: Penguin Classics, 2003).
Descartes, Rene. "Discourse on Method," in Roger Ariew and Eric Watkins, eds., *Modern Philosophy* (Indianapolis, IN: Hackett Publishing, 2009).
Devall, Bill. "The Deep Ecology Movement." *Natural Resources Journal*, Vol. 20, No. 2, 1980.
Devall, Bill and Geroge Sessions. *Deep Ecology: Living as if Nature Mattered* (Layton, Utah: Gibbs Smith, 2001).
Dreyfus, Hubert and Sean Dorrance Kelly. *All Things Shining: Reading the Western Classics to Find Meaning in a Secular Age* (New York: Free Press, 2011).
Durheim Emile. *The Division of Labor in Society* (New York: Free Press, 2014).
Ehrenriech, Barbara. *For Her Own Good: Two Centuries of Experts Advice to Women* (New York: Anchor, 2005).
Elias, Norbert. *The Society of Individuals* (London: Continuum, 2001).
Ewen, Stewart. *Captains of Consciousness: Advertising and the Social Roots of the Consumer Culture* (New York: Basic Books, 2001).
Feyerabend, Paul. *Against Method* (New York: Verso, 2010).

———. "How to Defend Society Against Science." *Radical Philosophy* Vol. 11, No. 1, 1975, pp. 3–9.

Freedberg, David. *The Power of Images: Studies in the History and Theory of Response* (Chicago: University of Chicago Press, 1991).

Foucault, Michel. *Discipline and Punish* (New York: Vintage Books, 1995).

———. *Power/Knowledge* (New York: Vintage, 1980).

———. *The History of Sexuality, Volume 1* (New York: Vintage Books, 1990).

Frankfurt, Henry G. *On Bullshit* (Princeton: Princeton University Press, 2005).

Furtak, Rick Anthony. *Wisdom in Love: Kierkegaard and the Ancient Quest for Emotional Integrity* (Notre Dame: University of Notre Dame Press, 2005).

Gadamer, Hans-Georg. *Truth and Method* (New York: Continuum, 1996).

Giddens, Anthony. *The Consequences of Modernity* (Stanford: Stanford University Press, 1991).

Gillespie, Michael Allen. *The Theological Origins of Modernity* (Chicago: University of Chicago Press, 2008).

Gross, Matthias and Linsey McGoey, eds. *Routledge International Handbook of Ignorance Studies* (New York: Routledge, 2015).

Habermas, Jürgen. *The Theory of Communicative Rationality, Vol. 1, Reason and the Rationalization of Society* (Boston, MA: Beacon Press, 1984).

Harding, Sandra, ed. *The Feminist Standpoint Theory Reader* (New York: Routledge, 2004).

Hegel, G. W. F. *The Phenomenology of Spirit* (New York: Oxford University Press, 1977).

Heidegger, Martin. *Being and Time* (London: SCM Press, 1962).

———. *The Question Concerning Technology and Other Essays* (New York: Harper, 1984).

Hénaff, Marcel. *The Price of Truth: Gift, Money and Philosophy* (Stanford, Stanford University Press, 2010).

Heshel, Abraham. *The Prophets* (New York: Perennial Classics, 2001).

Hochschild, Arlie Russel. *The Commercialization of Intimate Life: Notes from Work and Home* (Berkeley: University of California Press, 2003).

Honneth, Axel. *The Struggle for Recognition: The Moral Grammar of Social Conflicts* (Boston: The MIT Press, 1996).

Hooks, bel. *Teaching to Transgress: Education as the Practice of Freedom* (New York: Routledge, 1994).

Horheimer, Max and Theodor Adorno. *Dialectic of Enlightenment: Philosophical Fragments* (Stanford: Stanford University Press, 2002).

Hyde, Lewis. *The Gift: Imagination and the Erotic Life of Property* (New York: Vintage, 1983).

Jagger, Alison M. "Love and Knowledge: emotion in Feminist Epistemology." *Inquiry*, Vol. 32, No. 2, 1989, pp. 151–176.

Jaspers, Karl. *Man in the Modern Age* (New York: Doubleday, 1957).

Jonas, Hans. *The Gnostic Religion: The Message of the Alien God and the Beginnings of Christianity* (Boston: Beacon Press, 2001).

———. *The Imperative of Responsibility: In Search of an Ethics of a Technological Age* (Chicago: University Of Chicago Press, 1985).
———. *The Phenomenon of Life: Toward a Philosophical Biology* (Evanston, IL: Northwestern University Press, 2001).
Kittay, Eva Feder. *Love's Labor: Essays on Women, Equality and Dependency* (New York: Routledge, 1999).
Kittay, Eva Feder and Ellen K. Feder, eds. *The Subject of Care: Feminist Perspectives on Dependency* (Lanham, MD: Rowman and Littlefield, 2002).
Koyre, Alexandre. *From the Closed World to the Infinite Universe* (New York: Angelico Press, 2016).
Kuhn, Thomas. *The Structure of Scientific Revolutions* (Chicago: University of Chicago Press, 1996).
LaDuke, Winona. *Recovering the Sacred: The Power of Naming and Claiming* (Chicago: Haymarket Books, 2016).
Lakoff, George and Mark Johnson. *Metaphors We Live By* (Chicago: University of Chicago Press, 1980).
Leigh, McEvoy, Adrianne, ed. *Sex, Love and Friendship: Studies of the Society for the Philosophy of Sex and Love: 1993–2003* (Amsterdam: Rodopi, 1997).
Lloyd, Genevieve. "The Man of Reason." *Metaphilosophy*, Vol. 10, No. 1, January 1979.
McIntyre, Alasdair. *After Virtue* (Notre Dame, IN: Notre Dame Press, 2007).
Merchant, Carolyn. *The Death of Nature: Women Ecology and the Scientific Revolution* (New York: Harper and Row, 1990).
Midgley, Mary. *Science as Salvation: A Modern Myth and Its Meaning* (New York: Routledge, 1994).
———. *Wisdom, Information and Wonder: What is Knowledge For?* (New York: Routledge, 1989).
Mills, C. Wright. *The Sociological Imagination* (New York: Oxford University Press, 1959).
Mills, Charles W. *The Racial Contract* (Ithaca, NY: Cornell University Press, 1997).
Moruzzi, Norma. "From Parvenu to Pariah: Hannah Arendt's *Rahel Varnhagen*," in Samuel Fleischacker, ed., *Heidegger's Jewish Followers* (Pittsburgh, PA: Duquesne Press, 2008).
Mumford, Lewis. *The Myth of the Machine: The Pentagon of Power* (New York: Harcourt Brace Jovanovich, Inc., 1970).
Naess, Arnic. "The Shallow and the Deep, Long-Range Ecology Movement: A Summary." *Inquiry* Vol. 16, No. 1, 1973, pp. 95–100.
Nietzsche, Friedrich. "On the Future of Our Educational Institutions," in *The Complete Works of Friedrich Nietzsche, Vol. 3* (Edinburgh, UK: Morrison and Gibb Limited, 1910).
———. *The Birth of Tragedy* (New York: Dover, 1995).
———. *The Gay Science* (New York: Vintage, 1974).
———. *The Genealogy of Morals* (New York: Oxford University Press, 1999).
———. *The Will to Power* (New York: Vintage, 1968).
———. *Twilight of the Idols*, in *The Portable Nietzsche* (New York: Penguin, 1977).

Neiman, Susan. *Evil in Modern Thought: An Alternative History of Philosophy* (Princeton: Princeton University Press, 2004).

Nussbaum, Martha. *Love's Knowledge: Essays on Philosophy and Literature (*New York: Oxford University Press, 1992).

———. *The Fragility of Goodness: Luck and Ethics in Greek Tragedy and Philosophy* (New York: Cambridge University Press, 1989).

Pack, Justin. *How the Neoliberalization of Academia Leads to Thoughtlessness: Arendt and the Modern University* (New York: Lexington Books, 2018).

Pinker, Steven. *Enlightenment Now: The Case for Reason, Science, Humanism and Progress* (New York: Viking, 2018).

Pitkin, Hannah Fenichel. *The Attack of the Blob: Hannah Arendt's Concept of the Social* (Chicago: University of Chicago Press, 1998).

Plato. *Phaedrus* (Cambridge: Focus, 2003).

———. *The Republic*, trans. with commentary by R. E. Allen (New Haven, CT: Yale University Press, 2006).

———. *The Symposium*, trans. with commentary by R. E. Allen (New Haven, CT: Yale University Press, 1993).

Polanyi, Karl. *The Great Transformation: The Political and Economic Origins of Our Time* (Boston: Beacon Press, 2001).

Polanyi, Michael. *Personal Knowledge: Towards A Post-Critical Philosophy* (Chicago: Chicago University Press, 1974).

Putnam, Robert. *Bowling Alone: The Collapse and Revival of American Community* (New York: Touchstone Publishing, 2001).

Riesman, David. *The Lonely Crowd: A Study of the Changing American Character* (New Haven, CT: Yale University Press, 2001).

Rorty, Richard. *Philosophy and the Mirror of Nature* (Princeton: Princeton University Press, 1981).

Ruddick, Sarah. *Maternal Thinking: Towards a Politics of Peace* (Boston: Beacon Press, 1995).

Sahlins, Marshall. *Stone Age Economics* (Chicago: Aldine, 1994).

Salleh, Ariel. "Deeper the Deep Ecology: the Eco-Feminist Connection." *Environmental Ethics*, Vol. 6, No. 4, 1984.

Scharff, Robert C. and Cal Dusek, eds. *Philosophy of Technology* (Malden, MA: Wiley Blackwell, 2014).

Scott, James C. *Against the Grain: A Deep History of the Earliest States* (New Haven, CT: Yale University Press, 2017).

———. *Seeing Like A State: How Certain Schemes to Improve the Human Condition Have Failed* (New Haven, CT: Yale University Press, 1998).

———. *The Art of Not Being Governed: An Anarchist History of Upland Southeast Asia* (New Haven, CT: Yale University Press, 2010).

Seidman, Steven. *Romantic Longings: Love in America, 1830–1980* (New York: Routledge, 1993).

Sessions, George, ed. *Deep Ecology for the Twenty-First Century* (Shambhala, 1995).

Singer, Irving. *The Nature of Love, Vol. 1: Plato to Luther* (Chicago: University of Chicago Press, 1987).

———. *The Nature of Love: Courtly and Romantic* (Boston: The MIT Press, 2009).
———. *The Nature of Love: The Modern World* (Boston: The MIT Press, 2009).
Smith, Dorothy. *The Everyday World as Problematic: A Feminist Sociology* (Boston: Northeastern University Press, 1989).
Soble, Alan, ed. *Sex, Love and Friendship* (Amsterdam: Editions Rodopi, 1997).
Strasser, Susan. *Satisfaction Guaranteed: The Making of the American Mass Market* (Washington, DC: Smithsonian Institution Press, 1989).
Sturgeon Noël, ed. *Ecofeminist Natures* (New York: Routledge, 1997).
Sullivan, Shannon and Nancy Tuana, eds. *Race and Epistemologies of Ignorance* (Albany, NY: SUNY University Press, 2007).
Taminaux, Jacques. *The Thracian Maid and the Professional Thinker: Arendt and Heidegger* (Albany, NY: SUNY University Press, 1997).
Taylor, Charles. *A Secular Age* (Cambridge: Harvard University Press, 2007).
———. *Modern Social Imaginaries* (Durham, NC: Duke University Press, 2007).
———. *Philosophical Arguments* (Cambridge: Harvard University Press, 1997).
———. *Philosophical Papers: Volume 2, Philosophy and the Human Sciences* (New York, NY: Cambridge University Press, 1985).
———. *Sources of the Self: The Making of Modern Identity* (Cambridge: Harvard University Press, 1992).
———. *The Ethics of Authenticity* (Cambridge: Harvard University Press, 1991).
———. *The Language Animal: The Full Shape of the Human Linguistic Capacity* (Cambridge: The Belknap Press, 2016).
Tinker, George. "Religion", in Frederick Hoxie, ed. *Encyclopedia of North American Indians* (New York: Houghton, 1996).
Tönnies, Ferdinand. *Communities and Society* (New York: Dover Publishing, 2011).
Tucker, ed. *The Marx and Engels Reader (Second Edition)* (New York: W.W. Norton & Company, 1978).
Wall Kimmerer, Robin. *Braiding Sweetgrass: Indigenous Wisdom, Scientific Knowledge and the Teaching of Plants* (Minneapolis, MN: Milkweed Editions, 2015).
———. *Gathering Moss: A Natural and Cultural History of Mosses* (Corvallis, OR: Oregon State University Press, 2003).
Weber, Max. *Economy and Society* (Berkeley: University of California Press, 1978).
———. *The Protestant Ethic and the Spirit of Capitalism* (New York: Routledge, 2002).
Young-Bruehl, Elizabeth. *Hannah Arendt: For the Love of the World* (Haverford, CN: Yale University Press, 2004).
Young, Julian. *German Philosophy in the Twentieth Century* (New York: Routledge, 2018).
———. *Nietzsche's Philosophy of Religion* (New York: Cambridge University Press, 2006).
———. *The Death of God and the Meaning of Life* (New York: Routledge, 2014).

Index

About the Author

Justin Pack is a full-time lecturer at CSU Stanislaus where he teaches philosophy courses on technology, science fiction, environmentalism, and ethics. Indeed, we may say he follows the admonition of Arendt: to think what we are doing and to critique the outstanding characteristic of our time—thoughtlessness. He loves coastal redwoods, giant sequoias, birds, enchiladas verdes, ancient fortresses, and ruins.

www.ingramcontent.com/pod-product-compliance
Ingram Content Group UK Ltd.
Pitfield, Milton Keynes, MK11 3LW, UK
UKHW041628031026
15587UKWH00001B/137

9 781498 591348